# Why Geese Don't Get Obese [AND WE DO]

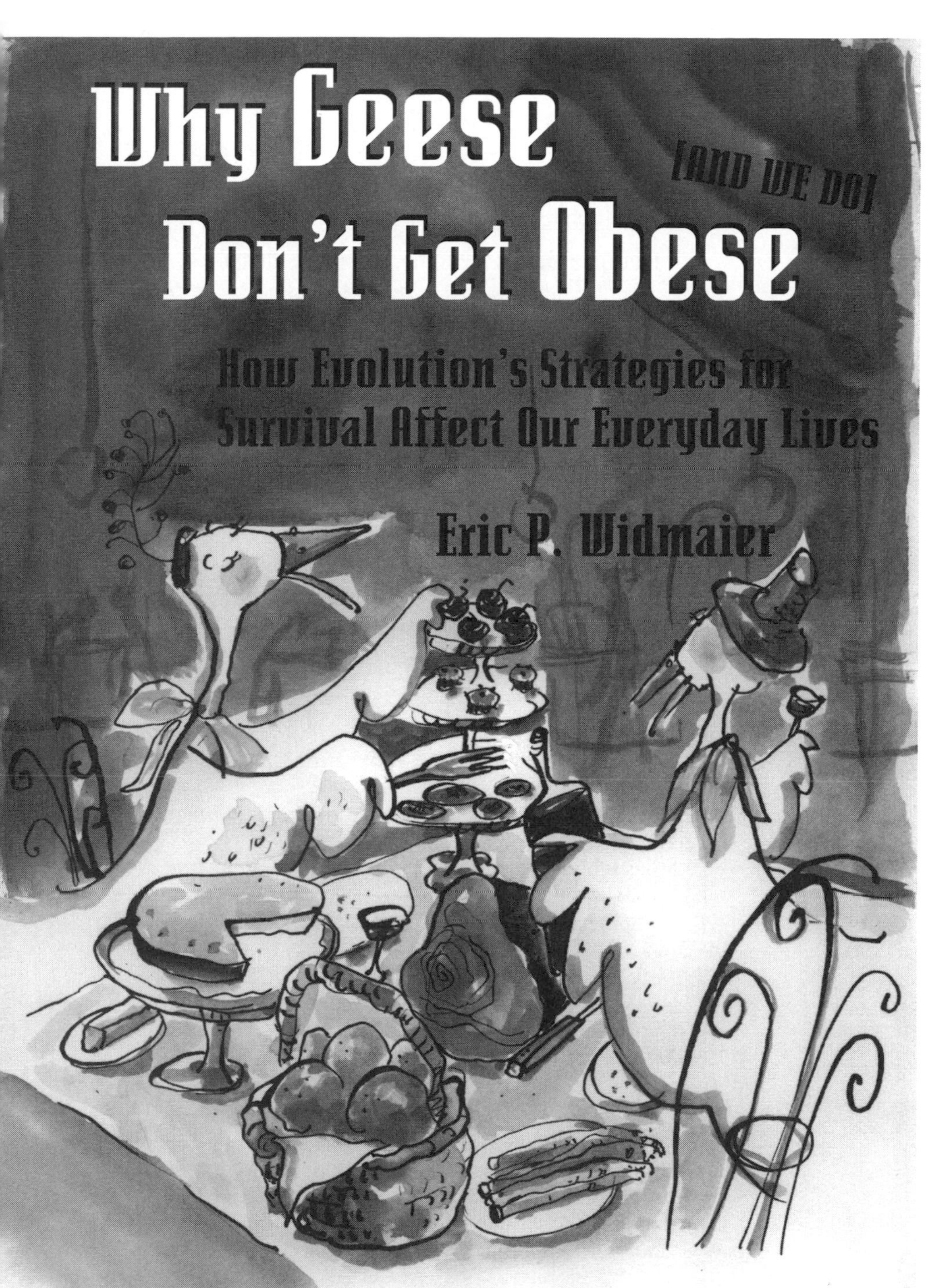


W. H. Freeman and Company • New York

This book is lovingly dedicated to Maria, Ricky, and Carrie, who give my life meaning, and to the memory of William and Mary Widmaier.

*Text designer: Victoria Tomaselli*

*Cover illustration and design: Janet Penderson*

Library of Congress Cataloging-in-Publication Data
Widmaier, Eric P.
Why geese don't get obese (and we do)/Eric P. Widmaier.
p. cm.
Includes bibliographical references and index.
ISBN 0-7167-3147-9
1. Physiology, Comparative—Popular works. I. Title
QP33.W53 1998
571.1—dc21 98-2698
CIP



Printed in the United States of America
First printing 1998

# CONTENTS

PREFACE vii

ACKNOWLEDGMENTS ix

CHAPTER ONE
Different Species, Same Problems 1

CHAPTER TWO
1,000 Cheeseburgers for Lunch, or Getting Enough to Eat 6

CHAPTER THREE
Too Much to Eat! 18

CHAPTER FOUR
Getting Enough to Drink (Water, That Is) 41

CHAPTER FIVE
Oxygen—The Breath of Life 53

CHAPTER SIX
Life Under Pressure 71

CHAPTER SEVEN
Bat Wings and Elephant Ears: Keeping Cool 91

CHAPTER EIGHT

Sensing the World Around Us 105

CHAPTER NINE

Stone Age Stress and Coping with Change 125

CHAPTER TEN

An Alternate Evolution 142

EPILOGUE

Doing Physiology 149

NOTES 156

INDEX 187

# PREFACE

The essence of physiology—questioning how and why the parts of the body function they way they do—probably began when ancient hominids first looked inside a dismembered animal and wondered what it all was. No doubt their real interest was whether or not they could eat any of the bits, but certainly there was a time when one of our early ancestors began to ask what all these entrails were for.

Although much has been learned, in a broad sense not much has changed about the nature of physiology since that early hominid. It is still a discipline intertwined with anatomy on every level, and a structure's form often provides important clues about its function. And while molecular biology is currently receiving a great deal of well-deserved attention (it seems we hear almost weekly about the discovery of some new disease-related gene), ultimately every major genetic discovery will need to be characterized and understood in a real-life setting.

I lecture on all aspects of physiology to a variety of audiences, and I always find the interrelatedness of animals to be endlessly fascinating. A bird flying over the Himalayas, a fish swimming in the tropics, a crustacean living in the deep ocean, and a person typing at a computer are very different animals in very different settings, yet all share the same biological needs and face similar challenges to survival. Every animal needs oxygen and a way to transport it within its body. Likewise, all animals must be able to sense changes in their environments, to cope with those changes, to find sources of energy to drive the chemical reactions in their bodies, and so on. One remarkable feature of these common needs for survival is how a species' environment dictates what measures its members must take to satisfy those needs. Surviving at an elevation of 17,000 feet in the Andes requires much different coping mechanisms than survival under the sea, but in both cases the limited factor is still oxygen.

In this book, I've collected some of the material that has best captured the imaginations of the varied audiences to whom I've lectured. My hope is that after reading the book, the reader will have a new appreciation for the astonishing ways our bodies are suited for survival and how we and all other animals are more closely related than might be imagined. One indisputable fact is that the more we understand about other animals, the more we also understand about ourselves.

## ACKNOWLEDGMENTS

I most sincerely and gratefully acknowledge the invaluable editorial help and advice of John Michel and the rest of the wonderful staff at W. H. Freeman and Company. Special thanks to Dr. Elizabeth Knoll, who helped me get started on this project and pointed me in the right direction. I am extremely grateful to those individuals and organizations who provided me with photographs: Bat Conservation International, Inc., in Austin, Texas; Francis Countway Medical Library; Dr. Thomas Eisner, Cornell University; Dr. Thomas H. Kunz, Boston University; and particularly Dr. Charles K. Levy, Boston University, for his photos, advice, and good cheer. I am also grateful to the National Institutes of Health and the National Science Foundation, which have supported my research on animal and human physiology for many years, and to Boston University for providing me with the opportunity to pursue my research and teaching interests. Most of all, I thank my wife, Maria—who is much more literate than I ever will be—for her editorial help.

## CHAPTER ONE

# Different Species, Same Problems

*Nature is an endless combination and repetition of a very few laws. She hums the old well-known air through innumerable variations.*

—RALPH WALDO EMERSON, *ESSAYS* (1841)

When I was an eager young student at Northwestern University, I had the good fortune to be taught by a physics instructor who took a great interest in his students' academic careers. Like a true physicist, he never could really appreciate my reasons for wanting to study biology. And while his many attempts to get me to switch from a career in the life sciences to one in the physical sciences were ultimately unsuccessful, he nonetheless left me with some advice that I continue to pass on to the young biology students that I now teach. That advice is as simple as it is fundamental: Never forget that the laws of nature are at the root of the life sciences. In other words, to understand how the human body works, you must first understand something about the physical laws of gravity, electromagnetism, thermodynamics, and matter and energy. For example, it's highly unlikely that Sir Isaac Newton was thinking about the way blood flows to the head of a giraffe when the apple dropped on his head. However, his revelation about gravity led not only to a better understanding of how planets revolve around the sun but also helps us to understand why people sometimes get light-headed when they stand up too suddenly or why a giraffe's blood pressure must be higher than our own.

Don't be alarmed. You won't need a Ph.D. in physics to understand how the forces of nature influence how your body works. In the following chapters we'll see how warm-blooded animals, like ourselves, use heat energy to their advantage, why we have two nostrils, why seals don't get the bends, how sharks use electricity to monitor their surroundings, how the salt content of water determines whether or not a fish will drink (it's the opposite of what you might think!), and why elephants have such big, floppy ears.

The business of studying how the different structures of our bodies—such as the heart, brain, kidneys, and muscles—function, is the science known as physiology. This branch of science may have gotten its name from the Greek *physiologoi*, which was the name given to an ancient group of well-to-do philosophers. One of their favorite occupations was to debate the principles of nature and how those principles could explain the nature of living things. Many of their conclusions may not have made sense by today's standards, but nonetheless physiology took hold as a science and is stronger than ever as we enter the twenty-first century.[1]

As is the case with the elephant's ears, it is a common theme in physiology that even the oddest-looking creature appears that way for a reason. In fact, many of our own features that we take for granted are, on the surface, sort of strange looking, too. Why do we have two nostrils, for example? Wouldn't it make more sense to have one large opening in the nose rather than two smaller ones? And speaking of things that come in pairs, why do we have two eyes and two ears but only one tongue, when all of these structures are used for sensing things in the environment? Why don't we have a forked tongue like snakes? Why is it that some people are skinny, and others cannot seem to keep weight off no matter how hard they try? Likewise, why don't small animals like mice and shrews—who eat their body weight in food each day—get fat? And why might the ability of humans to gain weight actually have been an evolutionary advantage, one that has gone haywire in the modern era of fast foods and sugary sweets? All of these questions and many others like them can be answered if we accept the premise that nearly every change in an animal's form arose because of evolutionary pressures and the need to adapt to the environment.

As animals evolved in splendid ways in response to their environments, the laws of nature often created previously nonexistent problems. When the giraffe's neck got longer, for example, the animal was better able to eat vegetation that other animals couldn't reach. That's an obvious advantage, but the long neck created a new problem—how could blood get all the way from the heart up to the brain, a distance of many feet? Gravity works against the blood, of course, making it hard for the fluid to move upward. It may not seem that gravity would pose that much of a problem, but try connecting several straws together and see how quickly it becomes more difficult to sip from a glass. Somehow the system manages to work, however, because giraffes are extremely successful animals and live long lives. In fact, in order to solve the problem of gravity and get blood all the way up to the head, nature made the blood pressure of the giraffe very high, much higher than our own—a simple enough solution. But we all know that high blood pressure is deadly in people. Are giraffes somehow resistant to the dangers of high blood pressure, and, if so, wouldn't it be nice to know why, so that we may someday apply that knowledge to the human condition?

It's good to keep in mind that all animals, no matter what type of environment they live in, face the same challenges of survival. For example, whether the environment is a desert or an ocean, the body's water stores must be kept at proper levels. Even fish need the right mechanisms to keep from becoming dehydrated. Similarly, a cave-dwelling bat in Malaysia, a llama or a person in the Andes, a fish in the Pacific, and a crab in a tidal pool must all obtain sufficient oxygen from their environment to power the chemical reactions of the cells in their bodies. They all need some sort of pressurized blood circulation to move the oxygen from place to place within their bodies. The ways in which a person and a fish get oxygen and transport it around their bodies, however, are determined by the environments in which they live. Thus, lungs would do a fish no good, and gills wouldn't help us. As another example, a shark living in deep, murky waters needs to know what objects are in its immediate vicinity, just as a rat that is active at night does. But eyes are almost useless in murky water or on a dark night, so sharks and rodents need to rely on other sensory cues to "see" their surroundings. Sharks developed the ability to detect even the tiniest electrical signals given off by prey even when the prey is "playing possum." Rodents solved the same problem by developing an enormously enlarged smell center in their brains, and even the faintest odor tells a rat or mouse all kinds of useful information about what's nearby.

The way in which we maintain relatively constant levels of salt, water, oxygen, and blood pressure is called homeostasis. We will revisit this concept in Chapter 9, but for now it's worth mentioning that homeostasis is the very basis of health. Disease, in fact, can be defined as a state of nonhomeostasis.

Think of it as a balance between opposing forces. If you eat an entire pepperoni pizza, the salt level in your blood may rise to potentially harmful levels. You will be in danger of falling out of homeostasis. Fortunately, there are hormonal, behavioral, and brain mechanisms that set into motion a chain of events that quickly bring the salt concentration in the blood back to normal, restoring the homeostatic state. We all need these and many other built-in homeostatic controls, or in a very short time we would succumb to the rigors of the external world.[2]

Thus, the physical laws of nature and the environment in which an animal lives, combine to produce the incredible (but understandable) variety of shapes, appearances, and behaviors found in the animal kingdom, and even in ourselves. Every species must develop survival strategies to cope with the same basic, fundamental challenges: getting enough to eat, drink, and breathe; circulating blood; adapting to change; keeping warm; and communicating with other members of the species (or with other species). As a good illustration of how these principles come together, try to imagine a warm-blooded mammal so tiny it is barely heavier than a large insect. What problems would this produce and how would those problems be solved? If we had to deal with those same problems, how would they affect us? In fact, such a mammal does exist. It's called a shrew, and, as we'll see in the next chapter, if humans shared the physiological characteristics of shrews and other small mammals we could not possibly exist.

CHAPTER TWO

# 1,000 Cheeseburgers for Lunch, or Getting Enough to Eat

*Viewed narrowly, all life is universal hunger and an expression of energy associated with it.*

—MARY RITTER BEARD, HISTORIAN AND SUFFRAGIST, IN *UNDERSTANDING WOMEN* (1931)

Have you ever noticed how small animals are constantly scurrying about for food? Small animals like birds and squirrels and mice seem to be always running around looking for something to eat. On the other hand, cows spend a good bit of their day grazing but don't seem to be in a hurry about it (and how filling can grass be?). Lions spend time on a hunt but they seem to spend a good deal of time napping as well. Could size and eating habits have anything to do with each other? They could and they do.

With the exception of breathing, perhaps the most basic need of all animals, including ourselves, is getting enough to eat. We need a nearly continuous infusion of fuel in the form of food to meet our energy needs. Were it not for our large stomachs, we, too, would need to constantly eat to support our metabolisms. Fortunately for us, however, our needs and those of a squirrel are quite different.

Most people consume approximately 2,000 (women) to 2,500 (men) calories each day. At the high end of the scale, a male athlete exercising to exhaustion for an entire day needs about 7,500 calories to keep up with his energy needs. Naturally, if we burn up as many calories as we eat, our body weight will remain fairly constant. But imagine that you could eat about 200,000 calories each day and never gain weight! It sounds impossible, but if your body chemistry was the same as that of the smallest mammals, like shrews, that's just about what it would take to sustain you. The reason has to do with metabolic rate, which is much more sluggish in us than it is in mice, shrews, and other little mammals. Let's take a look at the meaning of metabolic rate, and try to imagine how we could satisfy a need for so many calories each day if we were human-sized shrews.

Picture an enclosed, room-sized chamber with nothing in it but a chair—no lights, appliances, or any other objects that could give off heat. Next, imagine that the chamber sits inside another, slightly larger chamber filled with water. If you were to enter the inner chamber and sit quietly in the chair, heat from your body would enter the air and warm it up by a small but measurable amount. The heat would then pass through the walls of the inner chamber and into the surrounding water, thereby raising the temperature of the water ever so slightly. The degree to which the water temperature rose would be a measure of your basal metabolic rate. If you got up from the chair and began running in place, you would generate more heat, which would cause the water temperature to rise even more, because your metabolic rate had increased. The energy that powers heat production comes from the burning of calories, which is why a person with a high basal metabolic rate not only tends to feel warmer than others but also has an easier time keeping trim.

This scenario is similar to the way an animal's metabolic rate is normally measured. Because animals exhibit different levels of activity (for example, think of a sloth and a mouse), we usually use the basal metabolic rate to compare different species. For practical reasons, sometimes it's easier to use oxygen consumption to measure metabolic rate. Because oxygen is needed to burn the sugars and fats that we use as fuel, this method is a good marker of how actively we are burning up calories. The more calories we burn, the more oxygen we consume during respiration.[1]

But what is metabolic rate, and how is it defined? Metabolic rate is simply the sum total of all the chemical reactions occurring in the body at any one time. The higher the rate, the faster the overall rate of our countless chemical reactions. Thus, the higher the rate, the more fuel is burned and more oxygen is consumed. (When we exercise we breathe faster.) Obviously, this means that more fuel must be provided (more food consumed) to replenish the energy used.

Differences in metabolic rate can mean important physiological differences in our daily lives. For example, the married couple's argument on whether the bedroom windows should be open or closed at night is an age-old battle. Typically, the woman, whose metabolic rate is often lower than that of a man, wants the windows closed because she feels cold. The man, with his higher metabolic rate, feels warm and wants the windows open. Another obvious and often-cited example are people who can eat like a horse and never gain weight versus those who say they need only look at ice cream to put on a few pounds.

Even these simple examples give some idea of the diversity of metabolic rates among humans. Imagine then how difficult it is to compare metabolic rates between species, comparing, say, a dog to a cow, cat, or human.

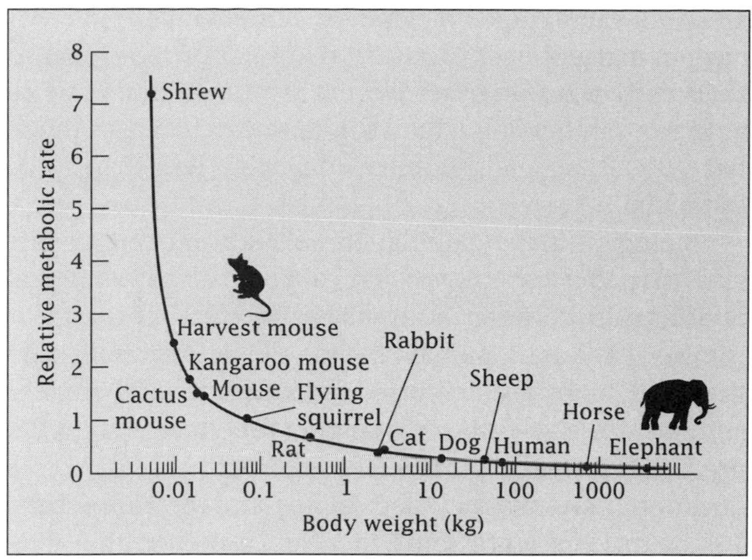


*As animal species increase in size, their metabolic rate decreases, but the relationship between size and metabolic rate is not a simple linear one, and at the extremes of body size, metabolic rate changes exponentially. Tiny animals, like shrews and mice, have extraordinarily high metabolic rates, and must eat constantly to replenish the energy used up to sustain their metabolism. (Modified from Randall et al., eds.,* Eckert: Animal Physiology: Mechanisms and Adaptations, *4th ed. Copyright © 1997 by W. H. Freeman and Company. Used with permission.)*

But by using devices like the water chamber or the oxygen-consumption idea, we can make fair estimates for each species. When we do so, we discover some very surprising things.

For the dozens of mammalian species that have been analyzed, a simple pattern has emerged between body size and metabolic rate—bigger species have lower metabolic rates than smaller species. This is all relative, of course. Common sense tells us that an elephant generates more heat, burns up more fuel, and consumes more oxygen than a mouse. When you compare the body sizes of a mouse and an elephant, however, you find that oxygen use and heat production are not proportional to their body sizes. In other words, an animal that is twice as big as another animal does not produce twice as much heat. It will, instead, produce somewhat less than twice as much.

Biologists have catalogued so many different species that they now recognize a precise mathematical relationship between body size and metabolic rate. It is clear that larger animals have a slower metabolic rate than one might predict based on their size, and smaller animals have a faster metabolic rate than might be predicted for their size. But why should this be? It's a question that has puzzled biologists for more than 100 years. There are some theories, however, that may explain why this pattern exists in nature.

Think of a newborn baby swaddled in blankets. All parents know that infants need to be warmly dressed, even at temperatures considered mild by adults. By the time this baby has grown to adulthood, it will have grown in height by a factor of about 3.5 (for example, a 20-inch-long baby who grows to 6 feet), but its body proportions won't grow at the same ratio. For example, a very young child's head makes up about twice as much of its total body length as does that of an adult. What's important here is that as the baby (or any animal) grows, its weight (mass) increases more than its body surface (the so-called surface area). Thus, although an adult has more total surface area on its body, a baby's surface area relative to its mass is actually greater.

What does surface area have to do with keeping a baby warm? It's the same as soup in a broad, flat bowl cooling faster than soup in a narrow cup. An adult is like the narrow cup—lots of mass but not much surface area. The baby is like the flat bowl—much more surface area relative to its mass. Therefore, the baby loses heat more readily than an adult, even though both are warm-blooded and have similar body temperatures.

Now we can see why a mouse or a shrew has a bigger problem maintaining its body temperature than an elephant. An elephant's body is so massive that it acts like a heat retainer, or "heat sink," as it's often called. Once a body that big is warmed up, it holds its heat for a long time. But a little creature, like a shrew or mouse, loses its heat very quickly. As heat is lost, it must be replaced, and the primary way heat is replaced is by burning more fuels like sugars and fats. That means every cell in the body must burn more fuel at a faster rate, or the body temperature will start falling. Each cell in a shrew's body is filled with more fuel-burning enzymes than a comparable cell in our body. Similarly, our body cells have more of these enzymes than the cells of an elephant.[2]

## Feeding Our Metabolisms

Little animals, therefore, must eat incredible amounts of food each day to provide the needed fuel to replace their heat losses. How much is an "incredible amount"? Imagine eating more than your own body weight each day and you'll get an idea of the challenge. Small mammals, like bats and shrews, actually consume up to one and a half times their body weight in food every day! For an adult man, that would be like eating about 1,000 quarter-pound cheeseburgers a day, every day of his life. Or, to put it another way, remember how full you feel after a huge Thanksgiving dinner. Now, imagine eating about 50 of those dinners each day, without gaining a single pound. This is how much food we would need to eat if our metabolism were that of a shrew, or if a shrew were to grow to our size but keep its same metabolism.

If all of this seems hard to imagine, it will be even more surprising to learn that shrews are not actually the smallest mammals. The bumblebee bat appears to have that distinction, weighing barely 2 grams, or less than one-tenth of an ounce. In fact, the bumblebee bat and the hummingbird may be the smallest warm-blooded creatures that ever lived or ever could live. Anything smaller could probably not produce enough heat to stay warm. Because of that, many small species of bats inhabit caves in temperate climates, where the cave temperature can reach 90 degrees or more. If a tiny bumblebee bat lived near the Arctic Circle, it would have to be covered with enough insulating fur to fill a basketball!

In short, any animal with a metabolism this great would need to be either eating or looking for food nearly every waking moment of its life. Big animals, on the other hand, have the luxury of free time to devote to

*The bumblebee bat, weighing less than a penny, may be the smallest warm-blooded mammal to ever live. If a warm-blooded animal were any smaller, it is doubtful that it could eat enough to maintain its body temperature. With such a high surface area-to-weight ratio, heat would be lost through the skin faster than it could be replaced by burning fuel.* *(Photo copyright ©Merlin D. Tuttle, Bat Conservation International. Used with permission.)*

other activities; they don't need a constant supply of food because they are not in danger of losing body heat so rapidly.

Now, if you were a shrew and had to constantly search for insects and other things to eat, how would you do it? Since a shrew is a tiny, weak animal, it is certainly not going to take a leisurely stroll outside its hiding place to look for a bite to eat. On the contrary, small animals tend to scurry and dart about, making it harder for owls, cats, and other hunters to catch them. Thus, not only do they need to eat a lot to provide fuel for heating their bodies, but they also need to replace the energy their muscles use for frenetic activity. In fact, it's hard to imagine how an animal like a shrew manages to survive at all.

If it is to have any chance of surviving, however, a shrew needs an efficient delivery system to process all that food and get the nutrients to the heart, brain, muscles, and other organs. Think about that Thanksgiving dinner. If you're like most people, the things you feel like doing after such a meal are sleeping, sitting, watching television, or otherwise taking it easy. The feeling of fullness lasts for hours. If you were like a shrew, however, and had another 49 turkey dinners waiting for you that day, you would have to get that food out of your stomach right away to make room for more. Thus, the digestive system of small animals is designed for fantastically quick breakdown and absorption of food. A large cockroach can be consumed by a shrew in moments and immediately processed by the digestive system. The nutrients are then rapidly shuttled into the blood, and room is made for the next meal.

Besides a huge food intake and an efficient digestive system, the shrew needs something else in order to make use of such a high metabolic rate. Food products like sugars and fatty acids can only be fully burned by the body's cells in the presence of oxygen (just as a fire in a fireplace burns brighter when you blow some air on it). To keep up with the high food intake, a shrew or other small animal must have sufficient oxygen available. There are two ways in which this can be done. First, a small animal might have unusually large lungs for its size so that each breath would take in more oxygen. However, it would be pretty hard to design an animal the size of a mouse with lungs the size of, say, a raccoon. In fact, we find that the size of an animal's lungs increases exactly in proportion to the size of the animal.[3]

The other way to get more oxygen is simply to breathe faster. Not surprisingly, then, smaller animals breathe at a faster rate than larger animals. For example, while people take about 12 (men) to 15 (women) breaths per minute while resting, the smallest mammals may take up to 100 breaths or more each minute, nearly two per second. (Try that for a moment—you'll find it impossible to sustain.) Thus, the tremendous amount of oxygen used by the shrew to burn all that fuel is replaced as soon as it's used.[4]

## Cardiac Performance

Once an animal has lots of fuel and sufficient oxygen to burn the fuel, somehow both the fuel and the oxygen must be delivered to every cell in the body. This is where the cardiovascular system of the mammal comes in. All mammals have a four-chambered heart that supplies blood to all parts of the body by pressure. Blood flows away from the heart in the high-pressure arteries and is returned through the low-pressure veins. As with the lungs (or any other organ), there are two ways to get more out of the heart—make it bigger or make it work (pump) faster. It turns out that, like the lungs, the size of an animal's heart is exactly proportional to the size of the animal. Thus, a four-ton elephant has a heart that is four times the size of a one-ton horse. Because the heart is a muscle, it can enlarge and become stronger. This is what happens in people when they do regular aerobic exercise, which creates a much more powerful heart. But it would be difficult to fit a huge heart inside a small animal's chest, and the heart within a given species will remain more or less within a strict size range.

Heart rates, on the other hand, do vary between species, and, in general, the smaller the animal, the faster the heart rate. Our hearts beat about 70 times per minute, while shrews, hummingbirds, and small species of bats have heart rates up to 600 beats per minute, or 10 beats per second! And the rate can double when the animals are active.

In order to see a better comparison between the human heart and that of a shrew, it is worth doing some simple arithmetic. If you multiply the heart rate by the amount of blood pumped out with each beat, you get a figure known as the "cardiac output." For example, each time our heart beats, it pumps out about one-twelfth of a quart of blood. When you consider our heart rate is about 70 beats per minute, multiplying this number by the amount of blood pumped with each beat gives you a cardiac output of five to six quarts per minute.

Each time a shrew's heart beats, only 10 to 20 *millionths* of a quart of blood is pumped out! Multiplying that value by a heart rate of 600 beats per minute, we find that its cardiac output is about one-eightieth of a quart per minute, compared to our own five to six quarts per minute—or about 400 times less. Because our body weight is about 40,000 times that of a shrew and the shrew's cardiac output is only 400 times less, its heart must be working about 100 times harder than ours. By having a faster heart rate, the circulatory system of shrews and other small animals captures more of the oxygen inhaled through the rapidly expanding lungs and keeps that blood flowing at a high rate to every part of the body.

Thus, nature has figured out that it's better to keep the internal organs of a tiny animal's body small, but let them work harder to keep up with the animal's metabolism. At the other end of the spectrum is the blue whale, the largest mammal in the world and possibly the largest animal ever. The heart of these magnificent animals weighs about 1,300 pounds, or roughly the size of a large cow. In comparison, our hearts weigh about three-quarters of a pound. With each beat, about 25 gallons of blood is pumped out of each side of a blue whale's heart, but because the metabolic rate of this giant is so low compared to small mammals, its heart rate is only about 10 or fewer beats per minute. Think of the power behind each beat, however, as 25 gallons of blood gushes out into the arteries. It just goes to show that everything is relative. A mouse's heart looks pretty puny to us, but our own heart is dwarfed by the whale's. In fact, approximately the same number of mouse hearts (1,000 or so) could fit into ours as the number of our hearts could fit into a blue whale's heart.

Given the mathematical relationships we've just explored, we can accurately predict an animal's heart rate, breathing rate, and metabolic rate if we know its body mass. Those relationships indicate that these three rates all vary with body mass in the same way. As stated earlier, the age-old explanation for this was the connection between heat loss and body size. This made such good sense, that most scientists were content with the solution for nearly 100 years. Unfortunately, it turns out that this can't really be the entire answer to why small animals have such high metabolic rates. The challenge to this received wisdom came when scientists realized that *all* animals, both warm-blooded and cold-blooded, show the identical relationship between metabolic rate and body size. Because cold-blooded animals do not produce their own heat and rely upon the sun for body heat, the idea of needing more fuel to replace heat lost through surface area cannot apply. When a cold-blooded animal loses heat, it merely basks in the sun until it gets its heat back again. There is no reason to assume that its metabolic rate would be tied up with how rapidly it loses internal heat.

Nonetheless, it is hard to imagine that all these tight mathematical correlations are mere coincidences. Somehow, it seems certain that body size, surface area, heat loss, and metabolism must be related. Just as certain, however, is that the relationship is probably much more complex than we presently realize. One of the wilder theories suggests that the numbers balance out better if the variable of time is included, that is, small animals live much shorter lives than larger animals. A rodent lives only about two years, if it's lucky, while some large animals live nearly 100 years. So, the thinking goes, all animals have a similar quota of energy to burn during their lives. Highly active, metabolic animals like hummingbirds and rodents use up as much energy in a year or two as a sluggish elephant does in 40 years. It's an interesting idea, but it's also without much scientific basis.

It would seem to be a pretty good thing that humans exist somewhere in the middle of the spectrum of metabolic rates. It means that we don't have to eat constantly, but we do have to eat at least a couple of times a day. We're not so huge as to become a heat sink, but not so small as to be in constant danger of losing body temperature. In the next chapter, we'll see what happens to our metabolism when food is not available (starvation) and when food is plentiful and more is eaten than is needed for good health (obesity).

## CHAPTER THREE

# Too Much to Eat!

*I can reason down or deny everything, except this perpetual Belly: feed he must and will, and I cannot make him respectable.*

—RALPH WALDO EMERSON, AMERICAN POET, *REPRESENTATIVE MEN* (1850)

It is all too apparent a truth that Mr. Emerson was not alone in his battle with an ever-expanding stomach. Most of us have tried to lose a few pounds at one time or another. When we do, we often find that the simple formula that matches food consumption with metabolism doesn't add up. It's not a uniquely human problem. Plenty of examples exist in nature of animals that fatten up now and then. What's different between us and those other species, however, is that they fatten up for a reason, such as to prepare for migration or hibernation, while we put on weight apparently for no good reason.

Prior to migration, for example, a goose will consume the equivalent of up to 25 percent of its body weight each day, accumulating large amounts of fat in the process. While the goose could rightly be called fattened during this time, it is not obese, at least not in the way we traditionally define the term. The difference is more than a semantic one. Once the geese begin their arduous migration, they often cover up to 600 miles in a single day. And although flying looks effortless for any bird, it isn't. Flying requires significant energy, especially when battling strong winds. By the time their migration is completed, most birds are at or even below their premigratory body weights. Thus, a period of fattening is followed by a period of intense exercise that results in restoration of normal body weight. In other words, flying birds almost never lay down permanent deposits of fat. People, on the other hand, almost invariably do. Once we begin to put on weight, it becomes a chronic, lifelong battle to keep it in check. Therefore, what is called fattening in a goose might be called obesity in a person. There are no negative physiological consequences of fattening in migratory birds, but there are profound consequences of obesity in people (premature death being one).

Geese and other birds are only one of many examples of the differences between fattening in animals and obesity in people. Hibernating bats throughout the world, including most of the United States, increase their body weight by as much as 50 percent prior to winter. If we did that each year, our arteries would become clogged with atherosclerotic plaques ("hardening of the arteries"), our blood pressure would probably rise to unhealthy levels, and we might even develop such diseases as diabetes. Although it is tempting to think that bats and other animals do not suffer these problems because they have more healthy diets than humans, this turns out not to be the case. Professor Thomas Kunz of Boston University and I recently determined that when bats put on this much weight their blood cholesterol rises to very high levels, and their intake of fat (by eating fatty insects) increases enormously. Despite this, their coronary arteries are strikingly clean and show no evidence of even a trace of atherosclerosis. In short, while they consume a high-fat diet, become much heavier, and develop high cholesterol levels—everything we are told not to do—they are nevertheless perfectly healthy. They live to a ripe old age without any cardiovascular consequences of their behavior. Why are bats (and many other animals) protected from the diseases we suffer under such conditions? We don't know the answer, but, like migrating geese, bats expend enormous amounts of energy each night foraging for food. It could be that one link between migrating birds and hibernating animals is the extreme exercise that makes up part of their lives.

Incidentally, geese and other animals *can* become obese—not just fat—if the conditions are right. It doesn't normally happen in nature but can occur when animals are domesticated. An overfed, underexercised house pet can easily gain excessive weight and suffer many of the same health consequences as people. The practice of eating goose liver, which began in central Europe during the Bronze Age and continues today is another example. In this particularly gruesome example, geese are force-fed through a metal funnel up to eight times per day for more than two weeks, until they become so grotesquely obese that they have difficulty breathing. At that time, the geese are killed, and their livers, which have by then accumulated massive fatty deposits, are processed into foie gras, a high-fat pâté much in demand by those who favor such things.

Like geese, bats, and other wild animals, our ancestors may have been free of the cardiovascular problems associated with obesity in modern society. Early man most likely had a much more physically demanding life than we do today. Simply collecting food could have meant foraging for miles each day. Thus, the ability to gain large amounts of weight may actually have been a survival mechanism, that is, it may have arisen as an important adaptation in primitive humans and survived to this day. Before I can convince you that overweight people are the remnants of a more highly evolved heritage, however, we need to step back and learn what is known about how body weight is controlled. This is a field of science that has been actively investigated for many decades, but it is only since 1994 that it has blossomed into a more intensely researched and far better understood area. The key event that occurred in that year was the isolation, or cloning, of a so-called "obesity gene." But let's not get ahead of ourselves.

First of all, why do we eat? We know from Chapter 2 why we *need* to eat, but what compels people and all animals to *want* to eat? Although I am decidedly carnivorous, I must confess that, on a certain level, there is something rather disgusting about chomping into the flesh of another animal. And when you consider that our ancient ancestors ate pretty much the way animals eat today, namely hacking off pieces of meat, viscera, and sinew and eating it raw and bloodied, it's even more unappetizing. Yet at one time the sight of a dismembered, bloody, recently killed boar must have made our mouths water with anticipation (not that much different from the sight of a T-bone steak on the grill). So, it's logical to assume that the drive to eat is primitive and so compelling that our ancient ancestors were willing to bite into just about anything for nourishment. Even in today's world, there are still people so undernourished and desperate for food that they will eat scraps of food from garbage dumps with little hesitation. As horrible as that is, it reinforces the notion that the drive to eat is overwhelming and beyond our control, except for limited amounts of time.

Whatever the biochemical basis of that drive may be, it's easy to imagine why it is necessary. With all of the other preoccupations of animals and primitive man, if there were no compelling force making them want to eat, they could easily forget about it and waste away.

But eating is not designed simply to put on calories and make us look nice and plump. It is also the way all animals get most of their vitamins and minerals, without which all sorts of nasty diseases arise, such as beriberi, scurvy, hypothyroidism, and pernicious anemia, all quite harmful. And for many animals, like those living in deserts, eating is the major source of vital water, because many foods (especially vegetation) are very high in water content.

It can even be argued that the urge to eat helped mold animal societies into what they are today. For instance, lions have a complex social arrangement within a troop, and much of it revolves around a cooperative effort to track and capture prey. Similar things could be said for all pack-hunting carnivores. Herbivores, which are the prey of carnivores, evolved a social structure called "herding," in which they congregate in large numbers for protection while grazing. By increasing their proximity to each other, herbivores developed social interactions designed to minimize hostilities within the group and protect the young from danger. Even early hominids were probably pushed along the social ladder when they began forming social units designed to increase food collection, make more effective hunting forays, and avoid becoming prey themselves. Now, let's try to decipher exactly what this drive to eat is and where it comes from.

## It All Begins in the Brain

Generally speaking, whenever you look for a control mechanism of a behavior, you're likely to find it within the brain. Eating is no exception. It turns out there are well-defined centers, called nuclei, within the brain that make us feel hungry or full.[1] However, the brain doesn't work on its own. It gets signals from the stomach, the eyes, the nose, and other regions. These signals may tell the brain that food is available, that the stomach is no longer stretched by food and therefore needs to be filled again, and so on. But how does the brain get information from these other areas of the body? In general, there are two ways in which the brain can communicate back and forth with other structures. The first is by neural signals that travel up and down the spinal cord and throughout the brain. The second is by hormones that are released into the blood and act on cells of the brain.

Let's take the case of the neural signals first. Scientists have long been aware that there are nuclei within a critical region of the brain called the hypothalamus, which regulates both feeding and metabolism. Although the hypothalamus is tiny and weighs less than one-seventh of an ounce, it is extremely important to a variety of functions. Located in the middle of the bottom of the brain, the hypothalamus contains many nuclei, which regulate such diverse activities as reproduction, sleep/wake rhythms, emotions, body temperature, hormone secretion, and feeding and metabolism. Right below it and connected by a thin strand of tissue is the pituitary gland. This is one of the chief hormone glands of the body and is the interface between the brain and the other hormone glands, such as the adrenals and thyroid.

Several lines of evidence have narrowed the feeding-related sites in the hypothalamus to a few specific nuclei, such as the ventromedial nuclei (VMN) and the lateral hypothalamic area (LHA). (The names refer to the geographical coordinates of where within the hypothalamus the nuclei reside.) We know, for example, that in lesioning experiments, if the VMN are obliterated by a jolt of electric current, animals overeat. In fact, a rat whose VMN have been destroyed will become so ridiculously obese that it won't be able to turn around in its cage. On the other hand, if the VMN are stimulated, eating will stop altogether. Thus, we might say that the VMN are a satiety center, and without the normal operation of the VMN neurons, an animal continues eating until it is almost ready to burst.

VMN-lesioned rats don't explode, of course, but they do get awfully ornery. In fact, their behavior changes remarkably from a docile laboratory rat to an aggressive animal that also shows changes in sexual behavior. This unexpected effect is a very good example of the difficulties in determining the specific functions of different brain regions. When you destroy an area like the VMN, you can't be sure other problems haven't been created as well. You may intend to simply eliminate a feeding control center but find out later that this part of the brain contributes to other behaviors as well. This is a common problem in neurobiology, and careful control experiments must be conducted to ensure that you are doing what you think you are doing and nothing else. Unfortunately, this is not always possible.

In general, there are only a few ways in which the physiological functions of different brain regions can be assessed. One is by the lesioning experiments described above for the VMN. Another technique uses the opposite approach. Rather than obliterating a brain region, you can electrically stimulate it and observe what happens. This has happened accidentally in human brain surgeries, where patients remain conscious during the procedures (the brain does not feel pain). For instance, as a surgeon lowers a probe into a diseased part of the brain, the patient may experience certain sensations, emotions, or movements, thus telling us something about the functions of that part of the brain in people. Sadly, one of the best ways in which we've learned about brain function in people is by examining the brains of people that have died from neurological disorders, such as Parkinson's disease or Alzheimer's disease.

By noting the regions of the brain of such individuals where damage appears to have occurred, we can surmise that those regions are responsible for normal functioning. Scientists can then test that hypothesis in animals by performing the lesioning and stimulation experiments. Once the hypothesis is confirmed, researchers can begin rational approaches to preventing or curing the disease in humans.

In addition to the VMN, the cells of the LHA also play a role in feeding, but this part of the brain is a feeding center. Stimulating this part of the hypothalamus causes an animal to start eating, even if it is not hungry. On the other hand, if you lesion the LHA, the animal won't eat. You might wonder what would happen if both the feeding and satiety centers were destroyed. It's a good question, and it tells us something very important about the dominance in these centers. It turns out that an animal treated in this way stops eating altogether, even though the satiety center is gone. Thus, the satiety center probably works by inhibiting the feeding center. If there is no longer a satiety center, the feeding center runs amok and animals overeat, but if the feeding center is also destroyed and there is nothing to instigate hunger in the first place, removing the satiety center makes no difference.

That is the basic mechanism, but as is so often the case with the brain, it's not as simple as it first appears. Other nuclei have recently been discovered to be vital links in the process as well. Each of these nuclei contain bundles of fibers that run back and forth between the other nuclei, thus setting up a sort of intrahypothalamic communication network. Signals that affect one nucleus, therefore, often affect the others. The back and forth signaling within the hypothalamus is actually a very common way by which different brain centers communicate. Neurons from one nucleus may receive and send inputs to many other sites. In fact, most neurons in the brain are interconnected with anywhere from hundreds to hundreds of thousands of other cells. This is one reason why the study of the brain can be so frustrating. Try to imagine a system containing roughly a trillion components (neurons), each of which is connected with thousands of other components. It's beyond our ability to comprehend. Even computer models of the brain are very difficult to establish; there is simply too much going on during every millisecond.[2]

Our understanding of the role of such complex brain chemistry and circuitry in the control of body weight has been greatly aided in the past 50 years by the discovery of several bizarre, mutant strains of mice and rats. Among these, the so-called fatty rat and the obese mouse have yielded the best clues about the mechanism of weight gain and metabolism. These strange rodents suffer from a variety of ailments, all of which can be traced back to their unfortunate tendency to become extremely obese. We usually think of the term "obese" in the context of human beings, but in the case of these two mutants, no other word could suffice. Like the VMN-lesioned rats, they become monstrously huge compared to their lean counterparts. Until recently the underlying cause of their obesity remained largely a mystery. About the only thing we did know prior to the recent discoveries is that one or another genes in the cells of these species had mutated in such a way as to cause two aberrations: first, the rodents ate too much; second, the metabolic rate of the animals slowed down. Although we don't "feel" it, this latter phenomenon is nevertheless familiar to anyone who has ever dieted—it is what causes the so-called yo-yo effect.

Many of us are familiar with how easy it is to lose a few pounds and how much harder it is to lose the next five or 10. That's because our bodies sense the decreased food intake and adjust to it by slowing down our metabolism (that is, how quickly we burn off calories). That may sound like a nasty trick of fate, but in fact it evolved as an adaptation to prolonged starvation. Thus, if our primitive ancestors were forced to go for days with little or no food, their bodies would try to accommodate for it by burning fuel at a slower rate. In much the same way, the brains of the obese rodents have somehow "decided" that the animals are starving when in reality just the opposite is true. Consequently, their metabolism is reduced to prevent the body's fuel stores from being depleted.

Putting yourself in the place of the first scientists who discovered these mutants, imagine how you would put the neuropeptide hypothesis to the test. Let's say, for example, that you have identified a peptide in the hypothalamus that you suspect is a major factor controlling appetite. Your hypothesis is that this peptide factor normally acts to increase the urge to eat. Thus, you might guess that mutant obese rodents would have more of this factor in their hypothalamus than lean rodents and then measure it. This is exactly what people predicted and what turned out to be true.

In fact, this is not only the case with rats but virtually with all mammals, including humans. Our brains are loaded with a peptide known as Neuropeptide Y (NPY). This peptide likely serves numerous critical biological functions but is especially important in regulating how much we eat. Mutant mice and rats have extra NPY in their hypothalamus, and if you inject NPY into the brain of a skinny rat, it will overeat and become obese. It may not come as a surprise that no one has yet volunteered to have NPY injected into his or her brain, but the inference is that it works the same way in people as it does in rodents and other animals. Thus, it would seem that if we can figure out why obese animals have too much NPY, we can permanently solve the obesity problem. If only things were that simple.

## Leptin: The Obesity Hormone

It turns out that one reason for the extra NPY in obese rodents is an inadequate amount of yet another peptide, a circulating hormone called leptin (from the Greek *leptos*, for thin). The existence of leptin was first postulated decades ago but was only proven in 1994, when the techniques of modern biology became sophisticated enough to detect it. The story of how people first got the idea that there may be a blood-borne factor regulating body weight is extraordinary. In the late 1960s and early 1970s, an investigator named Douglas Coleman of the Jackson Laboratory at Bar Harbor, Maine, began a series of surgical experiments on mice known as parabiosis experiments. In what may sound like science fiction, he surgically connected the blood systems of two different mice—one obese and one thin. In effect, he created a kind of Siamese twin mice, with only the bloodstreams connected to each other. By doing this, anything in the blood of one animal would eventually find its way into its "twin's" blood.

Coleman reasoned that obese mice ate too much and burned too few calories, because they were missing a factor in their blood that normally acted to reverse these processes. Of course, he knew nothing of the hormone that later became known as leptin, nor of the neuropeptides discussed in this chapter. But he figured that if normal mice have lots of this putative factor and mutant obese mice little or none, then by connecting them some of the factor should move from the skinny mouse's blood to its obese twin.

When he completed the experiments he found that the obese mice *did* get thinner and started to eat less than before. The skinny mice remained at their normal body weight. Here was the first proof that a blood-borne factor regulating body weight did indeed exist, and that obesity (at least in mutant mice) could be explained as the result of an imbalance in the level of that factor. It would be nearly 25 years before his idea was borne out by the sophisticated techniques of modern molecular biology.

Leptin is made in the fat cells of the body, known as adipocytes, and is the product of the so-called obese gene referred to at the beginning of this chapter. As a person eats more and puts on weight, he or she doesn't grow significant numbers of new adipocytes; rather, each individual adipocyte becomes larger as it fills up with triglycerides and other fats. The notion that the number of our adipocytes is relatively fixed during childhood has some interesting implications. As we age and, in many cases, put on weight, each individual fat cell becomes a bit fatter. Large numbers of new fat cells are not created—that happens primarily in early life. But if there were twice as many fat cells around to begin with, there would be that many more fat cells with the capacity to become fatter. So, the idea goes, it is especially important not to overfeed young children and toddlers, because this is the time when they can still make additional adipocytes. Overeating, therefore, will cause growth of both the number and the size of fat cells. It is highly probable that overeating in newborns leads to adults with lots of extra adipocytes, just waiting for that next cupcake or scoop of ice cream.

As adipocytes enlarge, however, they begin secreting more leptin. This is a good thing, because leptin acts in the brain to prevent the formation of NPY, which you'll recall is the neuropeptide that causes overeating. Not surprisingly, some mutant strains of obese rodents may produce an abnormal form of leptin—one that doesn't work as well as it should.

Thus, in a healthy, normal person who tries to lose 10 or so pounds by dieting, a familiar pattern emerges. As the first few pounds of weight are lost, it becomes increasingly difficult to lose more. This may be because as the fat cells shrink, less leptin is produced. Less leptin means less inhibition of NPY. As the amount of NPY rises in the brain, appetite is stimulated, making it harder and harder to resist the temptation to eat. Among dieters, this back-and-forth shift between weight loss due to reducing food or calorie intake and weight gain due to a slower metabolism often makes substantial weight loss difficult and sometimes creates a "yo-yo" effect. And if this rather dismal explanation weren't bad enough, it gets worse. NPY, it turns out, doesn't only stimulate appetite, it also slows down metabolism. That means that no matter what a person eats, the rate at which the body's cells burn calories slows down. The less calories you burn, the slower the weight comes off. So, as you lose weight, leptin levels fall, NPY content rises, appetite is stimulated, and body metabolism slows. It's a vicious cycle, to be sure, but one that arose to protect us from the effects of starvation when food sources were scarce or unpredictable rather than to thwart would-be dieters.

Or so the theory goes, anyway. It's important to bear in mind that most of these ideas come from careful animal experimentation and sophisticated molecular biology studies on animal cells. Can we really extrapolate these observations to human beings? The answer is a qualified yes. Qualified, because there is much more to the story than just NPY and leptin, and we know much less about the other molecules involved. For instance, there is also a sort of anti-NPY in the brain called corticotropin releasing–hormone, or CRH.[3] Too much of this hormone can make an animal, and presumably a person, anorexic. Yet another brain peptide, cholecystokinin (CCK), also reduces appetite if administered into the brain of obese animals. And the list of peptides continues to grow.

Clearly, the control of feeding and metabolism is of such profound and fundamental importance that the human body has evolved a fantastically complex control mechanism for its regulation. And someday it will become clear exactly how the dozens of factors involved in this process must interact to carefully regulate body weight. One theory that has gained wide acceptance is that every person has a built-in set point at which the body prefers to operate. A person who is extremely overweight may be at the mercy of this set point. The brain of such an individual, through the combined action of NPY, leptin, and other peptides, apparently determines that the obesity of such individuals is the desired normal body weight. This is analogous in some ways to a fever. During a fever, products from bacteria fool the brain into thinking that the normal set point of body temperature is greater than 98.6 degrees. No one knows for certain why this happens, because the higher temperatures provide a less hospitable environment for the bacteria. Nevertheless, the brain, thinking the body should really be, say, 101 degrees, compares its new set point (101) with the actual temperature (98.6) and decides the body is too cold. This initiates activities designed to warm us up, such as shivering (which is why we get the chills at the onset of a fever). The brain is trying to respond to its new thermostat setting.

Therefore, if a person is unlucky enough to be born with a brain that is programmed with a high set point for body weight, he or she will have a difficult time overcoming that programming. As this person loses weight, the brain will compare the actual body weight with the preprogrammed set point and incorrectly determine that he or she is becoming dangerously thin. This will cause the metabolism to slow down and the appetite to increase, and the familiar up-and-down cycle of weight control begins.

With the recent discoveries of leptin, CCK, CRH, and other molecules, we might imagine that increasing the amount of these peptides in the blood of an overweight individual would result in a decreased appetite and increased burning of calories. This may or may not prove to be true—it's too early to tell—but to show just how dangerous a little knowledge can be, consider what happens in the case of CCK.

For a very brief period health food stores began expounding CCK as a "natural" method of weight control. In theory, the idea sounds fine if you believe the animal studies. The problem, however, is that CCK acts on the brain, and our intestines cannot readily absorb large peptide molecules; they must first be broken into smaller chunks. So, when you ingest CCK in the form of a powder or pill, your intestines first break down the peptide into its building blocks, the amino acids. Once in the blood, the amino acids do not magically reform into the same peptide or protein that was swallowed. Instead, the amino acids are absorbed by the cells of the body and build up into whatever proteins those particular cells need at that time. Thus, taking CCK pills was an exercise in futility. The same argument would hold for leptin. In fact, this is why diabetics need to inject themselves with insulin, rather than simply take an "insulin pill," even if one existed. The only way that oral administration of CCK could work is if a chemically modified form of the peptide was developed that could be absorbed across the intestinal wall and then reach the relevant regions of the brain that control feeding. This may someday be possible and is being actively researched by at least one major pharmaceutical company.

What about injecting oneself with CCK or leptin? Even then it might not work. First, the animal studies indicate that these compounds work best if injected directly into the brain. Second, even if they do reach the brain after being injected systemically, there is still no guarantee of success. We might predict that obese individuals, for example, would have abnormally low levels of leptin in their blood. Such a situation might arise if, say, the gene for making leptin was missing in such individuals. Recent evidence, however, suggests that most obese people have not less but more leptin than thin people (although in June 1997 the first case of leptin deficiency in humans was reported). This finding, surprising though it may seem, was not entirely unexpected. We can understand just how this could be, if we look at what triggers the two major forms of another common problem in humans, particulary those who are overweight—diabetes mellitus.

## Two Types of Diabetes

The two major types of diabetes are differentiated by the mechanisms that cause each. In Type I diabetes, the part of the pancreas that makes insulin is destroyed and cannot regrow. Individuals with Type I diabetes must inject themselves with insulin every day for the rest of their lives, in order to replace the insulin lost from the diseased pancreas.[4] Fortunately, this type makes up only about 5 percent or so of all cases of diabetes, at least in the United States. Originally, Type I diabetes was known simply as diabetes or as juvenile diabetes, because it usually manifests itself during childhood. In fact, the word "juvenile" is not completely accurate because it also develops in many adults. The name "Type I" was adopted when it was recognized that there is a Type II form of the disease. Nowadays, Type I is also called insulin-dependent diabetes mellitus (IDDM).

It is now believed that IDDM is a member of a general class of diseases known as autoimmune diseases. These diseases, like lupus, multiple sclerosis, Addison's disease, myasthenia gravis, and many others, are characterized by an immune system gone haywire. For reasons that are still unclear, the immune system recognizes its own cells as being foreign and attacks them just as they might attack cells from a skin graft (in other words, the immune system rejects a person's own body). In IDDM, the part of the pancreas that makes insulin becomes the foreign body and is all too efficiently destroyed by the overactive immune system. Although current treatments for the disease depend upon strategies for replacing insulin with injections or other means, ultimately the cure for the disease will involve targeting the immune disorder before it has a chance to develop. Oddly, many people are apparently born with a predisposition for diabetes but never develop the disease. In identical twins, for instance, only about one-third of the cases where one twin has IDDM does the other have it as well. This number is too high to be explained by pure chance but too low to be purely genetic. Clearly, there are environmental triggers (possibly infections) that initiate the immune response in these individuals. Our best guess is that only if a person with the right genetic background is exposed to those triggers will he or she develop the disease.

In Type II diabetes, or non-insulin-dependent diabetes mellitus (NIDDM), the pancreas is functional, but, for reasons that are still unclear, the ability of the body's cells to respond to insulin has been lost or diminished. The symptoms of Type II diabetes usually do not occur until around midlife, and for that reason it is also known as adult-onset diabetes. In these cases, the pancreas is normal and makes insulin, but the hormone is not very effective. Thus, the symptoms of this disease are essentially the same as those of Type I diabetes. Interestingly, the blood insulin levels of such patients are often even higher than normal, as if the pancreas were trying to compensate for the lack of insulin's effectiveness by cranking out extra amounts of the hormone. Nonetheless, the treatment for such patients is to induce the healthy pancreas to make yet more insulin (usually with a pill taken orally once a day). The hope is that by swamping the cells with excess insulin they will respond better. This treatment generally works pretty well but insulin injections are occasionally needed.

Fortunately, exercise and weight loss make our body's cells more sensitive to insulin. In fact, many Type II diabetics require no medication at all once they modify their diet and incorporate exercise into their daily routine.

## Obesity and Its Consequences

So what of leptin and obese people? It appears that obesity in the human population is akin to the Type II diabetic model. An early clue that this might be the case arose from Douglas Coleman's work. In the course of his studies, he observed a variant of the obese mutant mice that did not get thin when connected to normal mice. He concluded (correctly) that this variant was resistant to the weight-reducing factor, and we now know that they are indeed resistant to leptin. Interestingly, in Coleman's experiments the skinny mice connected to the resistant mice stopped eating altogether and nearly starved to death. This led to the idea that resistant mice did possess the weight-reducing factor—probably in abnormally high amounts. Thus, when they were connected to the skinny mice, the excess factor moved from the obese to the thin mouse's circulation, and the skinny mouse's brain perceived that the weight loss factor was very high and thus eating stopped completely. This is very similar to the conditions of NIDDM and obesity in humans. In other words, leptin is available in the blood (in fact, in amounts larger than normal), but the brain cells that normally respond to leptin fail to do so.

As you might imagine, the cellular basis of this insensitivity is currently being rigorously investigated by numerous laboratories around the world and several major pharmaceutical companies. When you consider that roughly one-third or more of the U.S. population is overweight (at least 30 million adults, according to estimates of the National Institutes of Health), the potential market for weight-reducing drug treatments is enormous. For example, consider that the rights to the cloned gene for leptin were sold by Rockefeller University (where Dr. Jeffrey Friedman, the discoverer of the gene, worked) to the California biotech company AMGEN, Inc., for $20 million in 1995, shortly after the discovery of the gene. And that was before it was even clear whether or not leptin would act in humans the way it acted in mice. In fact, that's still not clear. Nonetheless, in May 1996 leptin officially entered phase I clinical trials.

## The Body Mass Index

Up to now, we've alluded to the dangers of chronic obesity but have yet to actually define what is meant by the term. Clinically, obesity is defined by the body mass index (BMI). The BMI is defined as your weight in kilograms divided by the square of your height in meters. To get your weight in kilograms, divide your number of pounds by 2.2. For your height in meters, divide your height in inches by 39.3. You can estimate your BMI from the accompanying table in which I've converted height and weight to the more familiar units of inches and pounds. A 200-pound man who is 5 feet, 9 inches tall, for example, would have a BMI of 29.5. Although no consensus has been reached on the precise cutoff, if your BMI is at or above 27, you are considered obese and at higher risk for the diseases associated with obesity. Many people who consider themselves only a few pounds overweight will be surprised to learn they fall into the category of obesity.

Using a value of 27 for BMI, it has been estimated by the U.S. Department of Health and Human Services that

### *The Body Mass Index*

| Height (inches) | Weight (pounds) 110 | 120 | 130 | 140 | 150 | 160 | 170 | 180 | 200 | 220 | 240 |
|---|---|---|---|---|---|---|---|---|---|---|---|
| **62** | 19.8 | 21.6 | 23.4 | 25.2 | 26.9 | 28.7 | 30.5 | 32.3 | 35.9 | 39.5 | 43.1 |
| **63** | 19.5 | 21.2 | 23.0 | 24.8 | 26.5 | 28.3 | 30.1 | 31.9 | 35.4 | 38.1 | 42.4 |
| **64** | 18.9 | 20.6 | 22.3 | 24.0 | 25.7 | 27.4 | 29.2 | 30.9 | 34.5 | 37.7 | 41.1 |
| **65** | 18.3 | 20.0 | 21.7 | 23.3 | 25.0 | 26.6 | 28.3 | 30.0 | 33.3 | 36.6 | 39.9 |
| **66** | 17.7 | 19.3 | 21.0 | 22.6 | 24.2 | 25.8 | 27.4 | 29.0 | 32.2 | 35.6 | 38.8 |
| **67** | 17.2 | 18.8 | 20.3 | 21.9 | 23.5 | 25.0 | 26.6 | 28.1 | 31.3 | 34.6 | 37.7 |
| **68** | 16.7 | 18.2 | 19.7 | 21.3 | 22.8 | 24.3 | 25.8 | 27.3 | 30.4 | 33.4 | 36.4 |
| **69** | 16.2 | 17.7 | 19.2 | 20.6 | 22.2 | 23.6 | 25.1 | 26.5 | 29.5 | 32.4 | 35.3 |
| **70** | 15.8 | 17.2 | 18.6 | 20.1 | 21.5 | 22.9 | 24.4 | 25.8 | 28.7 | 31.5 | 34.3 |
| **71** | 15.3 | 16.7 | 18.1 | 19.5 | 20.9 | 22.3 | 23.7 | 25.1 | 27.9 | 30.6 | 33.4 |
| **72** | 14.9 | 16.2 | 17.6 | 19.0 | 20.3 | 21.7 | 23.0 | 24.4 | 27.1 | 29.8 | 32.5 |
| **73** | 14.5 | 15.8 | 17.1 | 18.4 | 19.8 | 21.1 | 22.4 | 23.7 | 26.3 | 29.0 | 31.6 |
| **74** | 14.1 | 15.4 | 16.7 | 17.9 | 19.2 | 20.5 | 21.8 | 23.1 | 25.6 | 28.2 | 30.8 |

up to 35 percent of adult men and 40 percent of women in the United States are obese. And the percentages rise to nearly 40 percent and 49 percent, respectively, by age 65. Thereafter, as overall health declines, the percentage of obese people rapidly declines, too. Thirty years ago, the numbers were considerably better: about 22–28 percent of men and 24–43 percent of women, depending on age. The single biggest change over the past 30 years, however, has been in young women aged 20–34 years, in whom the percentage of obese people has increased from a respectable 13 percent in 1962 to 25 percent today. This is of great concern because the earlier obesity is established, the more likely it is that serious complications will develop.

What are some of the diseases associated with obesity? Diabetes, for one, and all the secondary complications of that disease (stroke, blindness, gangrene). Atherosclerosis, heart failure, and hypertension are others—all of them serious and life-threatening. The economic ramifications alone of obesity are phenomenal.

At this point, we must ask why differences in leptin sensitivity evolved in the first place. Are we perhaps looking at the problem from the wrong perspective (that is, a modern one)? As I indicated earlier, perhaps the tendency to put on weight was actually an evolutionary jump rather than an unfortunate genetic error. Can the ability to put on large amounts of weight ever be considered an advantage?

To answer this question, we need only look at some very familiar animals that typically put on great amounts of weight each year and yet are perfectly healthy. Before a bird that summers in the north can migrate south for the winter, it must lay down large stores of fat. This provides the fuel source it needs during its long and arduous journey. Likewise, marmots, ground squirrels, bats, and other hibernators eat more and deposit more fat just prior to hibernation. For them, being fat is not only a good thing, it is the only way to ensure survival during the long weeks or months without food. Some of the smaller migrating and hibernating species put on a really impressive amount of weight. The common little brown bat (*Myotis lucifugus*) found in New England weighs only about 9 grams (one-third of an ounce), but increases to about 13 grams before hibernation. Four grams may not sound like much, but on a human scale, that would be equivalent to an increase in body weight from 150 pounds in summer to 217 pounds in autumn, year after year after year.

Could this logic be applied to people? Why not? I imagine that our primitive ancestors were as likely to go without food for short periods as they were to be well fed. Mild or even serious starvation, especially during droughts or ice ages, would have been a common event. Thus, a person whose metabolism and set point were programmed to burn up fewer calories would tend to accumulate more fat than someone with a high metabolism. The fatter person would be better able to withstand periods of starvation than the thin one. We might say that having a brain designed to increase body fat content was an advantage for any animal, including *Homo sapiens*.

Nowadays, the same built-in programming that allowed ancient man to withstand the stress of starvation has run amok. Ancient man didn't have the temptations or even the abundance of foods available today. There were no cookies in bakery windows, no potato chips to snack on while watching TV, no grocery stores with all the food you could ever want. He ate what he caught or found lying around. And he worked hard for his food, which meant he used up precious calories simply finding something to eat (remember the poor shrew scurrying about). How many calories do you suppose we consume walking from the couch to the fridge?

So, here's a case of an evolutionary adaptation that has outlived its usefulness, at least in developed countries. Certainly, however, as much of the world's population continues to starve, the business of adjusting body metabolism to a set point is still vitally important. It has often been said that one of the great tragedies of human social evolution is that half the world's population worries about the consequences of overeating while the other half starves.

> *Let the stoics [that is, the scientists] say what they please, we do not eat for the good of living, but because the meat is savory and the appetite is keen.*
>
> —EMERSON, 1844

CHAPTER FOUR

# Getting Enough to Drink (Water, That Is)

*Water, water everywhere, nor any drop to drink.*

—SAMUEL TAYLOR COLERIDGE,
*THE RIME OF THE ANCIENT MARINER* (1798)

When was the last time you overindulged on something particularly salty—like a pepperoni pizza? Most likely, all that salt made you really thirsty, possibly for hours afterward. Drinking lots of water (or worse, soft drinks, which tend to contain some sodium) probably made you feel bloated. But even worse, the salt content of your bloodstream increased to high levels. Doing this once in a while is no big deal, but doing it every day, well, that's another story.

What is it about salt that is so bad? The answer lies within our hearts and brains, our two most important—and delicate—organs. It turns out that it only takes a very small change in the salt content of blood—only a few percent—to throw our nerve cells out of whack and disrupt heart function. It's the sodium part of sodium chloride (table salt) that does the damage, by causing heart and brain cells to fire erratically. Furthermore, the addition of salt to our blood tends to draw water from our cells in an effort to return the salt concentration of the blood to normal. Much like a garden slug sprinkled with salt, when our cells become dehydrated they shrivel up and die. This may not be very important if a few skin cells dry up, for instance, but if our brain cells die, it is disastrous, because, unlike other parts of the body, cells in the brain do not regrow.

There are three compartments in the body that contain water: the heart and blood vessels (which contain the plasma); the interior of the cells (intracellular fluid); and the spaces between the cells and the blood vessels (interstitial fluid). About 8 percent of all the water in the body is found in the blood, 25 percent is in the interstitial fluid, and the remaining 67 percent is found within our cells. All together, water makes up about 60 percent of our body weight.

If our blood becomes salty after a meal, the equilibrium between the three water compartments is altered. The first thing that happens is that water moves by osmosis from the interstitial fluid into the blood, in an attempt to neutralize the excess salt in the blood. (Osmosis is the movement of water from a less salty region to a more salty one. This occurs spontaneously anytime two solutions of water with different amounts of salt in them come into contact.) Although the osmosis tends to normalize the plasma salt level, the interstitial fluid loses some of its water, leaving behind a saltier fluid. This, in turn, draws water by osmosis from the intracellular fluid into the interstitial fluid. The salt concentration in the interstitial space is returned toward normal, but now the intracellular fluid has lost some of its water, and thus become saltier. With less fluid inside, the cells start to shrivel and shrink as they become more and more dehydrated. Shriveled cells do not function very well, and in the case of the brain this shape change can kill neurons. Many of us, whether we realize it or not,

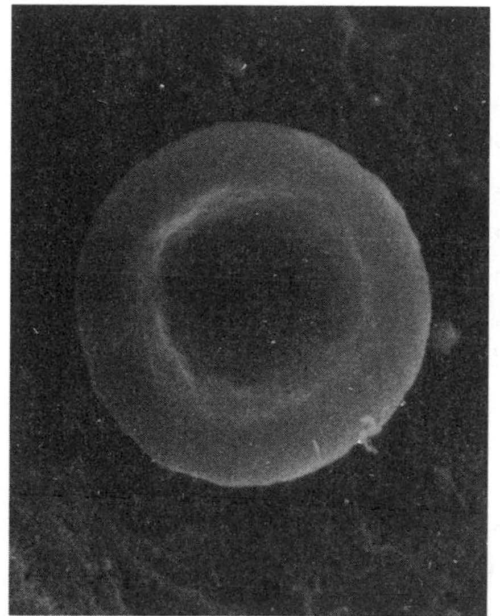

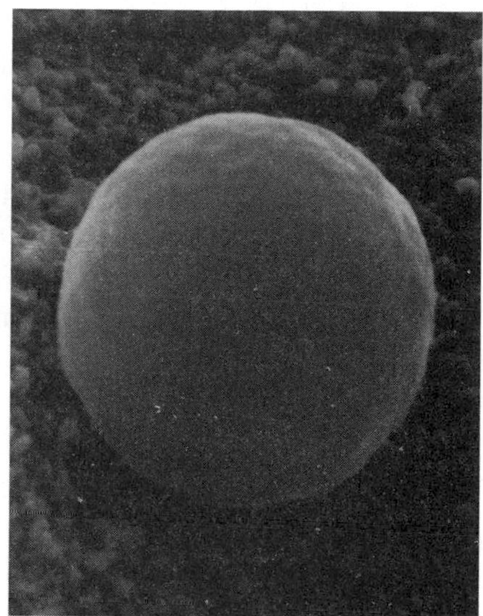

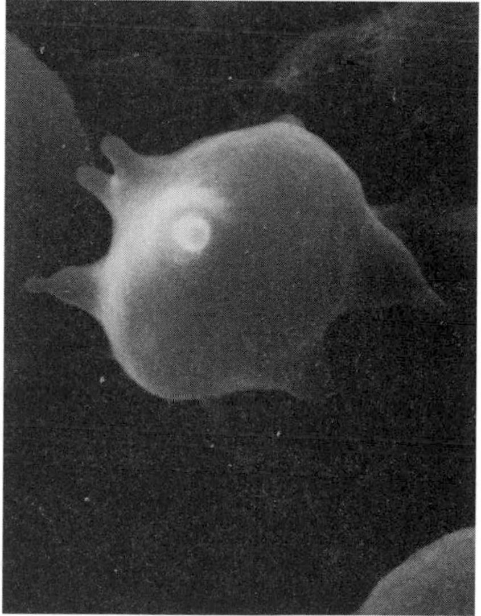

*In these remarkable scanning electron micrographs of individual red blood cells, the effects of the osmotic movement of water across cell membranes is vividly illustrated. From left to right, blood cells were immersed in a bathing fluid with a salt concentration that was the same, lower, and higher, respectively, than that of normal blood. Normal red blood cells have a concave shape but shrink in salty water and expand in diluted water. Were this to happen to the cells in our blood or any other part of the body (such as brain cells), the results would be disastrous. A shrunken or swollen cell loses its ability to function and, as in the case of red blood cells, can eventually rupture and die.* *(Micrographs copyright © David M. Phillips/Visuals Unlimited. Used with permission.)*

tend to be mildly dehydrated much of the time. We don't usually drink as much water as we should each day. (Incidentally, because alcohol also causes cells to become dehydrated but by a different mechanism, one way to prevent a hangover is to drink a glass or two of water after each alcohol-containing drink, thus preventing dehydration. Not surprisingly, the most common symptom of a hangover is the same as that of mild dehydration—a headache.)[1]

What else does excess salt do to the brain and other tissues? The brain and heart are examples of excitable tissues, that is, they are made of cells that change their characteristics when electrically stimulated. As early as the eighteenth century, Italian physiologists were aware of something they called "animal electricity." They knew that if the leg muscle of a frog were dissected with its major nerves intact, the muscle would jerk whenever two dissimilar metal rods were connected with the nerves. They deduced that some sort of charge went from the metal rods through the nerve and into the muscle, causing it to contract. Not a bad guess, really. We now know that nerves can generate their own electric charge. Unlike the current that flows through a wire, however, the current flowing through nerves is carried by electrolytes, such as sodium, potassium, and calcium.

If sodium or potassium levels in the fluid-filled spaces around cells become abnormally high, the nerves become unstable and are likely to fire off an electrical impulse even when they are unstimulated. These impulses are called action potentials, and they normally serve to transmit information from one nerve (or heart) cell to another. However, this system only works well if information is sent at the appropriate times. If impulses fire off randomly, the entire signaling process of the heart and brain becomes erratic and uncoordinated. As you might imagine, this can have lethal consequences. If the heart cells, for example, are getting impulses at the wrong times, the heart may beat before it has a chance to fill with blood, a condition that can be rapidly fatal.

Not surprisingly, there are numerous mechanisms in place to maintain a sodium, potassium, and water balance in virtually all animals, ourselves included. As long as we drink enough water and stay away from too many french fries and pepperoni pizzas, brain and heart cell shrinkage is usually not a problem for terrestrial animals like ourselves. Of course, animals living in very arid regions must constantly struggle to obtain sufficient water. Strangely, such species can exist almost entirely on the water their own bodies produce as a result of metabolism. (Water is one of the waste products generated in the body's cells when fuels are burned for energy.) Most of the water humans produce in this way is eliminated as water vapor each time we exhale. The rest of it is lost through the kidneys in urine. Desert animals, however, can recapture that water before it is lost.[2]

A good way to appreciate the need to balance salt and water in our bodies is to examine the problems encountered by animals that have to live in the stuff. For a fish in the ocean, the problem of keeping proper salt and water levels in the blood is much greater than it is for animals like ourselves. And unlikely though it may seem, the problem stems not from a need for either salt or water, but for oxygen.

## Getting Oxygen from Water

What does oxygen have to do with salt and water balance? Like all animals, fish need oxygen to survive. Oxygen helps provide the energy needed to drive chemical reactions, many of which are the same in fish and people. Essentially, all fish must get the oxygen they need from their only available source, the surrounding water. With a few unusual exceptions—known as lungfish[3]—saltwater and freshwater fish don't have lungs and must use their gills to breathe. The gills are hidden from view under a protective covering called the operculum, which fans out from a fish's body. Each time it does, it creates a flow of water that passes from the mouth over the gills and back into the ocean. Valves within the mouth of the fish make sure the water is directed out over the gills and not down the esophagus (the tube that connects the mouth to the stomach). To get water to move from the mouth to the operculum, a fish must create two chambers of negative pressure. Think of negative pressure as a sort of vacuum, where the water pressure inside the fish's mouth is lower than that in the ocean. All a fish needs to do is lower its bottom jaw, thus expanding the volume of its mouth. This lowers the water pressure in the mouth, and water moves from the ocean into the mouth, and not vice versa. To get water to continue its journey from the mouth to the gills, the opercula fan out from the body surface and create yet another vacuum-like chamber. This one is even more effective than the mouth vacuum pump, and water moves from the region of higher pressure (the mouth) to the region of lower pressure (the opercula). Finally, the opercula relax and close again, and the cycle starts over.[4]

Special cells within the gills capture oxygen from the water and deliver it into the bloodstream of the fish. Even though the gills are relatively small, they have a huge amount of surface area. Take for example the gills of a one pound trout—a moderately active fish. If unfolded and flattened out, the total surface of a trout's gills would cover a region of about one square yard.[5] In comparison, our own lungs unfolded in this way would cover a tennis court! Gills are loaded with extremely tiny blood vessels called capillaries, which deliver oxygen from the water to the bloodstream.

Incidentally, all those microscopic capillaries really add up when one considers the total area they cover. As an example, for each pound of weight we put on, we grow a mile of capillaries to supply the needs of the extra fat.

The gills of fish are thus ideally suited for oxygen capture and delivery. In fact, a fish is better able to remove oxygen from water than we are to get oxygen from air. The big drawback to water breathing, however, is that there is much less oxygen in water than in air. Fish must draw water over their gills at a very high rate to make up for the limited amount of oxygen dissolved in the ocean. Imagine if we had to breathe three times faster than normal all the time, and you get an idea of how hard fish work to get their oxygen.[6] All of this work takes energy and leaves less energy available for the other activities of fish. And, because they are cold-blooded, fish have less energy available than warm-blooded animals like us. But the danger of water loss is even worse for saltwater fish than the energy loss.

Unfortunately for fish, the very properties that make their gills ideal for capturing oxygen also make them perfectly suited for transporting water, because all thin, highly vascular structures like gills allow water to pass through their surfaces (that is, into or out of the blood vessels) by osmosis. Remember that in osmosis the saltless water moves toward the salty water. Thus, in saltwater fish, water from their blood (less salty) leaks out across the gills into the more salty surroundings of the environment, the seawater. It's bad enough the fish has to work so hard for its oxygen, but the loss of body water that results from osmosis over the gills presents some real problems.

A loss of water has two life-threatening consequences that would be the same in a person as in a fish. First, if the water content of an animal's body drops enough, it becomes impossible to provide sufficient blood pressure to send blood to all parts of the body. After all, blood is made up of about 50 percent water. No animal's heart is strong enough to pump blood to all the far reaches of its body if the blood level drops below a certain point. Second, the loss of body water means that the fluid left behind in the blood has a higher concentration of salt than normal.[7]

What about ourselves—do we ever lose water but retain salt? We do whenever we perspire. Although perspiration on our upper lip may taste salty, it is actually less salty than our blood. The water in the perspiration comes from our internal stores of water. Thus, when we perspire heavily on a hot day, we lose both water and salt but proportionately more water than salt. The fluid left behind in our blood, therefore, has a higher-than-normal salt concentration, just like a beaker of salt water that was allowed to evaporate. That's why the best way to recover from exercise on a hot summer day is to drink a solution that contains water and a little bit of salts dissolved in it, to exactly match what was lost through perspiration. In fact, many of the common sports drinks available today are really sugar-sweetened versions of (clean) human sweat![8]

## Seawater: To Drink or Not to Drink?

So how does a fish keep from losing water by osmosis through its gills? It would seem that the very process of "breathing" water has the unwanted side effect of producing a dangerously salty bloodstream as water is lost in the gills. Are fish defenseless against this?

Osmosis cannot be prevented, because it is a physical property of water that occurs automatically. Therefore, if fish are to survive, the water that's lost from the blood must be replaced. If a person were to become dehydrated, the cure would be simple—find a source of water and drink as much as needed. But ocean fish can't leave the ocean, so the only choice left is to drink the sea water. As strange as it may seem, an ocean fish must drink this salty fluid or its internal water level will drop, and its internal salt concentration will rise.

But drinking seawater presents a new problem—getting the water absorbed from the alimentary canal (the esophagus, stomach, and intestines) in the first place. Logically, you'd assume from the principle of osmosis that water would be drawn from the bloodstream (less salty) into the intestines, which have become filled with ingested salty water, but that is exactly the opposite of what a fish needs to happen! A fish is like any animal and needs to remove water from the intestines and add it to the blood. If water went the other way, it would only deplete the blood's stores even further (which is one reason why you don't drink sea water if you're stranded in the ocean).

To prevent such a disaster, an efficient mechanism evolved in marine fish to reverse osmosis. First, salt is removed from the alimentary canal by a set of molecular pumps and secreted into the fluid spaces outside the canal. This makes the fluid outside the intestines saltier than the fluid inside. Once the outside of the intestines is saltier than the inside, water moves automatically to the saltier side by osmosis. The water can then be carried off by the bloodstream. Thus, incredibly, the only way to get the water into the body is by first absorbing the salt with those molecular pumps. Without this mechanism, seawater could never get into the blood because the force of osmosis would be going the wrong way.

In short, saltwater fish need to drink seawater to replace the water they lose over their gills because of breathing. To absorb the water into their blood, they first need to absorb the salt, and water follows by osmosis. This solves the water-loss problem that was originally created by the need for oxygen. Or does it?

## What to Do with the Extra Salt?

Each time nature evolves a way to cope with one problem, it seems a new problem arises, one just as bad as the first. In the case of seawater fish, it's not obvious how this cycle can ever be broken. Surprisingly, however, the same structures that started this whole dilemma—the gills—are ultimately responsible for correcting it. The blood, with its load of salt water absorbed from the intestines, is pumped by the heart to the gills. On the gill surfaces yet another set of special cells with molecular pumps becomes active. These cells latch onto salt molecules in the blood and dump them out over the gills and back into the ocean. The water that remains behind is now free of its excess salt. Thus, each time a seawater fish takes a small drink of water, it ends up with a net increase in water, without salt, in its blood.

This may seem to be a fairly complicated way to balance salt and water levels in the body, but it also illustrates a common theme in nature—nothing comes for free. To get oxygen, fish evolved gills, but, as we have seen, the price is high. The form of payment comes in energy. All of those pumps that work on salt transport need a great deal of molecular energy (that is, ATP) for proper function. Energy used up for gill breathing is energy that is no longer available for other functions, such as muscle contraction. If a fish needs to escape from a predator, even a small decrease in available energy could result in its death.

Equally surprising is the freshwater fish, whose internal water (like our own) is much saltier than the surrounding lake or pond. By reversing the above reasoning, it should be clear that freshwater fish never drink water. In fact, freshwater fish must use energy to rid themselves of the tremendous amount of excess water coming in over their gills by osmosis. So much water enters the blood via the gills in freshwater fish that they have developed a very efficient pair of kidneys.

Picture them like a one-way faucet, always turned on full blast. If people had kidneys that worked in the same way, our bladders would be nearly bursting every 15 minutes—day and night!

Do other animals face similar challenges in water and salt balance? They do, and some more so than others. Oceangoing birds like seagulls inevitably ingest some salt when they feed on seawater fish. Like fish, these birds also have special salt-pumping cells (in pits below their eyes) to excrete the salt but retain the water. Certain reptiles, like marine iguanas and crocodiles, have similar salt pumping cells.

Thus, we see that all animals, large and small, aquatic or terrestrial, must cope with the problem of salt and water balance. But how about ourselves? If we eat a daily diet consisting of pizza, cheeseburgers, french fries, potato chips, and other salty foods, we are going to have a chronic excess of salt in our blood. An increase in blood salt of only about 1 percent will cause us to feel thirsty and, like all mammals, will produce a powerful urge to drink. We actually feel thirsty at two separate times after eating something salty. First, salt-detecting cells in the mouth, throat, and esophagus make us feel thirsty even before the salt has been absorbed by the intestine. This is a sort of fail-safe mechanism to make sure that water and salt are ingested together. Otherwise, the salt level would rise in the blood before water could be drunk to equalize it. Later on, after an hour or so, the brain senses the level of salt that has appeared in the blood after all the food and water has been absorbed from the intestines. At this point, if the amount of water consumed while eating was not quite enough to exactly match the salt concentration, the fine-tuning of salt levels takes place, as the brain perceives the imbalance and makes us thirsty again. The next time you fill up on fries and chips, think about it; you'll find you tend to feel thirsty again an hour or two later.

Drinking water will fix the salt imbalance brought about by our diet and protect our nerve cells from dehydration. As a result of drinking all that fluid, however, we become bloated, and this risks raising our blood pressure. Anytime fluid is added to a closed container—like our bloodstream—the pressure of the fluid increases. Unfortunately, we do not have special salt-pumping mechanisms like fish, marine reptiles, and marine birds, and thus we tend to suffer the consequences more than these animals.

Is it possible to suffer from too little salt in our diets? It is, but today most of us are so indoctrinated about the evils of salt in our diet that in some cases we reduce our salt intake too much. Just as too much sodium can alter cellular function, so can too little sodium (called hyponatremia). Excitable cells—those of the brain, heart, and muscles—need a finely balanced sodium concentration inside and outside their membranes, or their activities will become irregular.

It's clear that the evolution of life in an aquatic environment presented serious challenges that continue to plague animals like ourselves to this day. The mechanisms that developed to cope with these challenges may seem roundabout at first glance but in fact are entirely logical once we gain an understanding of the form and function of the organs and cells involved in the process. It is safe to say from this example that nature is not always predictable but, once understood, is always logical.

CHAPTER FIVE

# Oxygen—The Breath of Life

*I have discovered an air that is five or six times as good as common air.*

—JOSEPH PRIESTLEY,[1] ENGLISH CHEMIST (1774)

Have you ever heard a stubborn child in the midst of a tantrum shout, "I'll hold my breath until I turn blue!"? If so, you needn't worry, because try as he might, he won't be able to do it. Humans simply don't have the capability to go without air for very long. Similarly, most of us can only swim underwater for a minute or two before we're forced to come to the surface for air. Why can't we stay down longer? Why can't that stubborn child hold his breath until he passes out? It has to do with oxygen—the "good air" referred to by Priestley. Because oxygen makes up only one-fifth of all the gas molecules in air and, thus, one-fifth of the gas taken into our lungs, it doesn't take long to use up that meager supply in a single breath. As soon as that supply runs out, our body cells rapidly begin dying. Worse yet, our brain cells are among the first to go.

But how does our body know that the amount of oxygen in the blood is falling to dangerously low levels? And why can't we override the impulse to breathe?

Just as is true for food, salt, and water levels, there are special receptor cells in the body that determine when oxygen stores are too low. These cells, known as chemoreceptors, are located in large arteries, such as the carotid arteries. The carotids carry blood from the heart up to the brain. Because brain cells are more sensitive to changes in oxygen than any other cells, it is logical that oxygen-detector cells are located in arteries of the neck, rather than in arteries leading to your fingertips, for example.[2]

The arterial chemoreceptors communicate with the brain through a network of nerves. If oxygen levels fall, these nerves send signals to a primitive part of the brain called the brain stem (the region that connects the brain to the spinal cord).[3] Within the brain stem is a group of nerves collectively known as the respiratory center. When the center is stimulated by chemoreceptors, impulses travel from the brain stem and down the spinal cord to the rib muscles and to the diaphragm,[4] a large muscle at the bottom of the chest. When these muscles are stimulated by the respiratory nerves, they forcefully contract. It is the contraction of these muscles that makes our chest expand and take in air.

By breathing harder, more oxygen is added to the blood. Equally important, more carbon dioxide ($CO_2$), a waste product, is exhaled. Although we think of $CO_2$ as a harmless compound in the air, it can be fatal if allowed to accumulate in the blood. $CO_2$ can get into the plasma membranes that surround cells, especially nerve cells, disrupting their function; at high levels, it acts as a sort of anesthetic and can even desensitize the chemoreceptor cells. $CO_2$ also does not dissolve well in blood. As a consequence, we would fizz like a can of soda pop every time we moved around if we couldn't convert $CO_2$ into another form. Fortunately, there is an enzyme in red blood cells that converts $CO_2$ into harmless molecules of bicarbonate. Unfortunately, however, as $CO_2$ is converted to bicarbonate, it also generates a molecule of acid as a by-product, which is very harmful. High $CO_2$ levels, therefore, can lead to acidic blood, which in turn can wreak all kinds of havoc. It is not uncommon to see this pattern emerge in people with such respiratory diseases as emphysema.

Thus, we normally can't hold our breath long enough to pass out, because our chemoreceptors would override our conscious effort to stop breathing. This is true not only of humans but of every air-breathing animal. In fact, smaller animals tend to have even less ability to go without air. Many small rodents, for example, take up to 100 breaths per minute (compared to 12 to 15 in people) and could not withstand the loss of oxygen for even a few seconds. As a rule of thumb, warm-blooded animals, like us, tend to have higher rates of breathing, while cold-blooded animals, like reptiles, may take as little as one or two breaths per minute, a reflection of the higher metabolism of warm-blooded animals. More oxygen is required to provide the energy necessary for a mammal's higher metabolic rate.

## Marine Mammals—An Extreme Case

Some animals tolerate oxygen deficiency better than others. While most human beings die within five minutes of oxygen deprivation, marine mammals can go for much longer without air. This ability probably evolved as the terrestrial ancestors of whales and other marine mammals first started to spend longer and longer amounts of time in the water. Today's species, like certain seals, can survive well over an hour without taking a single breath. Exactly how they manage to do this has fascinated scientists for generations. One of the best-studied models of a diving marine mammal is the Weddell seal, which lives off the coasts of Antarctica.

This remarkable animal can repeatedly dive to great depths over the course of a day, sometimes staying underwater for more than 75 minutes (dolphins and whales generally can't stay under that long) and spending most its time foraging for food. A logical explanation for this unusual ability would be incredibly large lungs for the animal's size, which would allow it to take a massive breath before diving. This, however, turns out not to be true. In fact, some species of seals have smaller lungs than would be predicted for their body size. As a rule, however, lung size is perfectly proportional to body size.

Most diving mammals do, however, have more blood in their circulation than other animals of similar size. Circulating within the blood are the red blood cells, and within red blood cells is the protein hemoglobin. Hemoglobin, which is found in all vertebrates, contains iron, and it's the iron that gives hemoglobin its ability to bind oxygen molecules. That's why hemoglobin is known as the oxygen carrier of the blood (and, incidentally, is a major reason you need iron in your diet).

The more hemoglobin you have, the greater your oxygen reserve. A hormone made by the liver and kidneys called erythropoietin is the body's defense against anemia or low blood oxygen. This hormone, often called Epo for short, stimulates the bone marrow to produce more red blood cells. Animals and people that live at high altitudes have greater amounts of erythropoietin than sea-level dwellers. The extra red blood cells give them the ability to carry around more oxygen in the rarefied air (the higher the altitude, the less oxygen pressure there is). Elite athletes have known about this for years and have been known to engage in the controversial (and dangerous) activity of blood doping to enhance their performances. In blood doping, erythropoietin is repeatedly injected into an athlete. This increases the amount of red blood cells and, therefore, oxygen that can be carried in the blood. During the course of a race, even a little more oxygen in your blood may make the difference between winning and losing.

Although erythropoietin is only available for a small number of clinically approved uses, its use as a performance-enhancing drug in aerobic sports is widely accepted in the competitive cycling communities around the world. In the 1990's alone, dozens of European cyclists in the prime of life, and apparently in peak physical fitness, abruptly died from what appeared to be heart failure, but at least half of the deaths have defied clinical explanation. If an athlete (or anyone, for that matter) had been injecting himself or herself with erythropoietin, his or her blood would have become so overloaded with red blood cells that it would have turned to sludge. This would create a tremendous strain on the heart. If it were then subjected to something like the Tour de France, well, you can see how this might cause the heart to become overwhelmed.

The problem with Epo is that the only way to detect it is with a blood test, which is much more difficult to perform and more costly than a simple urine test, which is used to detect anabolic steroids. Also, the drug works best when used weeks ahead of a competition, because it takes time for the new red blood cells to be recruited from the bone marrow. That means the drug can leave the circulation after having done its job and is no longer detectable by racetime.[5]

Unlike blood dopers, seals have more hemoglobin in their blood for two reasons. First, they simply have more blood to begin with, even for animals their size. Second, they can "inject" red blood cells into their blood from a storage compartment in the spleen[6] exactly at the moment they begin diving. The signal for the spleen to empty its cells into the circulation is the hormone epinephrine. Incidentally, diving mammals aren't the only animals with this special type of contractile spleen; dogs, horses and many others have one, too. Why some animals have this capability and others don't remains a mystery. There is even some evidence that elite human divers, like female pearl divers in Korea, have a slight ability to spontaneously raise their blood cell count by this method. One great thing about such a system is that the increased blood cells are only present when they are needed—during a dive. When a seal surfaces, the blood cells are once again scooped up by the spleen and stored for the next dive. Thus, diving animals have their own natural version of blood doping that works just as well but without the dangerous long-term consequences.

This extra blood with its hemoglobin allows a seal about 15 minutes of additional underwater time but doesn't account for the remaining 45–60 minutes. Scientists are still debating what happens during this time, but a few possibilities have emerged. First, the muscles of seals are loaded with the protein myoglobin, which resembles hemoglobin in some ways, in particular because it can sequester oxygen from the blood. Myoglobin thus acts as a final reservoir of oxygen for the muscles. Once the oxygen in the blood has fallen to very low levels, the oxygen bound to myoglobin begins to come off and enter the cytoplasm of the muscle cell. Because a large proportion of a seal's total weight is muscle, much of the seal's body will get an extra 10 minutes or so of oxygen from myoglobin stores.

This still leaves about 30 or more unaccounted minutes of underwater time. Is there any molecule besides oxygen in the blood that can serve the same functions as oxygen? Not really. Even though only one-fifth of air is made up of oxygen, the remaining gas molecules (nitrogen, $CO_2$) turn out to be useless in the process of burning fuels to generate energy. So, that is clearly not the answer.

If oxygen is gone and no other gas in the blood can take its place, then there can be only one answer to the riddle of how a seal maintains energy stores underwater for so long. Somehow, energy must be produced in the absence of oxygen—a process called anaerobic respiration. All animals, including ourselves, have some ability to function by anaerobic respiration. There are, however, a few problems associated with this process. First, it is not an efficient way to produce the energy needed for cells to survive. Much more energy is formed when a single molecule of sugar is burned using oxygen than by anaerobic respiration. During exercise, our need for energy—and therefore oxygen—increases. During strenuous exercise, we can't always meet those demands and must rely partly upon anaerobic respiration. Second, one of the waste products of anaerobic respiration is lactic acid. Anyone who has ever strenuously exercised and experienced a leaden feeling in their arms knows exactly what lactic acid feels like in the blood.

Why is lactic acid so bad? As its name implies, it makes the blood more acidic. Changes in acidity, or pH, of the blood can be almost as devastating to survival as a lack of oxygen.[7] Therefore, how does a seal manage to function for more than half an hour on essentially 100 percent anaerobic respiration, especially if it repeats this process over and over again throughout the day? Part of the answer lies with their remarkable ability to alter their circulation during a dive, so that the amount of blood flowing to some parts of the body is maintained, while being decreased to other parts. Strangely, the large muscles are one of the regions that experiences a decrease in blood flow. At first glance, this may seem odd because the muscles are needed for swimming. Consider, however, that, unlike ourselves, a seal is ideally suited to swimming underwater. It has a beautifully streamlined body that makes swimming almost effortless. The muscles don't need as much blood and oxygen as you might imagine. As a consequence, the lactic acid that builds up in the muscles tends to remain there, rather than leaking into the entire circulation. This is beneficial, because it spares the heart and other structures from the potentially damaging effects of high lactic acid concentrations. When a seal eventually surfaces to breathe, the flow of blood into and out of the muscles slowly increases, and the lactic acid is quickly converted by oxygen into pyruvic acid, a harmless (useful, in fact) chemical.

If blood flowing to the muscles is temporarily slowed down during an anaerobic dive, where does the rest of the blood go? Much of it goes to the brain, which can never be deprived of blood—in a seal or any mammal. The heart, too, needs its oxygen. In pregnant seals, the placenta retains its normal blood flow during a dive, and the fetus's oxygen supply is not interrupted. Finally, the spleen (which holds the extra red blood cells), the retina (so the seal can retain sharp vision while searching for food), and the adrenal glands (which make the epinephrine that stimulates the spleen) all maintain their normal blood flow.

Thus, seals—and certain other mammals—have extraordinary abilities to go without oxygen for extended periods of time. For human beings, however, the limit is less than five minutes; a complete lack of oxygen for that long is inevitably associated with severe brain damage or death. In fact, one reason scientists study diving mammals is to hopefully learn more about the function of the brain and other structures at low oxygen levels and apply this information to improved therapies for stroke patients. Moreover, understanding why the internal organs of a seal do not suffer damage from oxygen deprivation may improve our ability to preserve human organs for transplantations, as well as prevent damage in shock victims.

## The Bends

When a human diver breathes air under water from a tank, the air must first be pressurized to overcome the hydrostatic pressure around the diver, that is, the increasing pressure of the water on the diver's body as she descends. Using pressurized gas makes it easier to fill and expand the lungs, but this creates a potentially dangerous problem of its own. To understand this problem and its effects, it is important to recognize that gases, like solids, dissolve in water. Just as table salt can dissolve in a cup of water, so can oxygen, $CO_2$, and nitrogen, the major components of air. And since our blood is roughly 50 percent water, gases can dissolve in our blood, too.

Normally, there is a limit to how much gas can dissolve in a fixed volume of water. One way to increase this limit, however, is by pressurizing the gas. A familiar example is the pressurized $CO_2$ dissolved in soft drinks and champagne. Breathing air under pressure is thus similar in some ways to adding carbonation to soft drinks. But what happens when a soft drink is shaken before being opened or when the cork is popped on a bottle of champagne? The pressure in the bottle is suddenly released, and the carbonated drink bubbles and fizzes. The bubbles are the gas molecules that have been jarred from the solution.

Likewise, if a diver has more than a normal amount of gas dissolved in her blood from breathing pressurized air, she runs the risk of exploding like a bottle of champagne when the pressure is released. This can happen when she ascends back to the surface and starts breathing air again at normal barometric pressure. The result can be a painful and potentially fatal illness known as decompression sickness, or the bends. What happens is that nitrogen gas, which makes up almost 80 percent of air and is not used in any way by the body, bubbles out of the blood and interferes with normal circulation. The only treatment for this is to repressurize the person in a hyperbaric chamber and force the nitrogen back into solution. This is followed by a painstakingly slow depressurization over many hours so that the nitrogen gas leaves the body in small, measured amounts. Because nitrogen can also have narcotic-like effects on human brain cells, the bends are also sometimes referred to as nitrogen narcosis.

Naturally, this phenomenon would happen anytime gas dissolved in the blood at one pressure were suddenly exposed to a lower pressure. Thus, you could get the bends if you were to be rapidly transported from sea level to high atop a mountain, where the air pressure is much lower. This can happen, for example, in unpressurized planes and was not unheard of among pilots during World War II.

Presumably, any natural diver like a seal must have evolved mechanisms to avoid—or at least reduce—the likelihood of getting the bends. It is hard to otherwise imagine such animals surviving, because they make countless dives over their lifetimes. And some of those dives are not only long in duration but reach unbelievable depths. Some whales, for instance, have had dives charted to depths of several thousand feet below the ocean surface.

So why don't seals and other diving mammals get the bends each time they dive? There is probably no single answer to this question. In fact, different diving mammals appear to solve the problem in different ways. Look again at the seal. Just before it enters the water, a seal does a curious thing. It exhales as completely as possible, which is the opposite of what we do before going underwater. But in this regard, the seal can be considered to be a bit smarter than us. Recall that the air in any mammal's lungs provides only enough oxygen for about two minutes of underwater time. Compared to the 75 or so minutes a seal can stay submerged, two extra minutes is pretty trivial. By exhaling, this two minutes worth of air is sacrificed, but the payoff is significant. The absence of inhaled air in the lungs means there is no air available to enter the bloodstream, and if no air is added to the bloodstream, there is no nitrogen gas to worry about when returning to the surface. Just to be sure, a seal's blood supply is redirected in such a way that much of it bypasses the lungs during its trip through the body. Thus, even if a little air were still in the lungs, it would not readily reach the bloodstream.

This is a remarkably simple solution to a complex problem. If you are a seal and want to avoid the bends, simply let out as much air as possible before diving and allow your circulatory system to do the rest. As simple and effective as this is, however, it is apparently not the only answer. Dolphins, for example, which are also good underwater divers, do not exhale first. And whether whales exhale or inhale before diving is uncertain (imagine trying to find out). On the other hand, the lungs of all diving mammals collapse due to the high water pressure around their chests. The air in the lungs is pushed up into the bronchi and trachea, where it cannot cross into the bloodstream. The unusual ease with which such animals reinflate their collapsed lungs upon surfacing may someday prove to be relevant to scientists looking for ways to treat people with collapsed lungs or chronic pulmonary diseases.

Scientists have proposed an intriguing hypothesis to explain why nitrogen gas doesn't bubble out of a dolphin's blood when it comes up from a dive. Think again of that can of soda pop or the bottle of champagne. When the lid is opened, carbon dioxide bubbles out. But if the can or bottle is shaken first, there is a much more violent reaction. Anytime a gas is dissolved under high pressure, it bubbles vigorously if shaken or disturbed.

In our bloodstream, blood doesn't move smoothly from place to place. On the contrary, it gets jostled, swirled, and squeezed as it moves from one blood vessel to another. As blood rushes under pressure from one artery to another, it often needs to make sharp turns, which causes lots of turbulence. In addition, the lining of our blood vessels is anything but smooth. If you were to look at the inside of one of our arteries, you'd see lots of nooks and crannies, jutting bits of debris, fatty plaques, and other features that give the blood a rough ride. In other words, our blood gets the same kind of treatment as the soda pop in the shaken can. Fortunately, gas doesn't normally start to fizz in our blood because there is only a small of amount of each gas present. But if the gas we breathed were to become pressurized, as in a dolphin that has just dived to a great depth, we would expect it to start bubbling.

Since large bubbles don't occur in the blood of dolphins, the theory has been put forward that the flow of blood in dolphins and possibly other diving mammals is smoother than in our bodies. The valves within the heart, for example, may open and close more smoothly. There are also very few of the fatty plaques so common in our arteries, and the angles that connect different blood vessels are less likely to cause turbulence. Thus, it may be that dolphins are simply "built" differently and that their form has perfectly evolved to suit their environment.

## Life in the Mountains

If sea mammals represent one type of adaptation to an extreme environment, what about the other extreme, where oxygen content is low? How do people and animals survive atop mountains and high plateaus, for example, where the air is thin?

Whether you're a llama, mountain goat, or a person working in the Andes Mountains, at an altitude, say, of 17,000–19,000 feet, the air you breathe is going to have a lower pressure of oxygen than the air I'm now breathing in Boston, but the percentage of oxygen is the same. Air is approximately 21 percent oxygen and the rest is primarily nitrogen, no matter where you are. The higher up you travel however, the lower the pressure of air.

The reason for this change is that just as water pressure increases around our bodies the deeper we dive, so does the air pressure increase the closer we are to the gravitational pull of the earth. You can picture what this means by imagining that you are standing at the bottom of a tall column of air that extends to the top of the atmosphere. At the bottom of the column (that is, sea level), there is more pressure than at the top, just as there would be more pressure at the bottom of the sea than at the top. (The only difference is that we can easily feel water pressure.) Thus, the barometric pressure in Denver (5,280 feet above sea level) is much lower than that of New York City (55 feet above sea level), and both are much higher than atop the Andes. What is the effect of less air pressure? The less pressure, the harder it is to drive air into the lungs and then force it across the lungs into the blood capillaries. So, if an animal or a person is going to live at high altitudes, they must develop better ways to capture and hold more of the oxygen that is available.

If you were to analyze the body shape and composition of a Peruvian mountain dweller, you'd find some striking differences from someone who has lived all his or her life at sea level. For one thing, the mountain dweller would be somewhat short, have a barrel-chested appearance, have more red blood cells, more blood in the circulation, and more blood vessels in his lungs.

What are the effects of these attributes? Short stature and an expanded chest volume means that people living at high altitudes have better lung capacity for their size. Furthermore, the increased amount of blood cells in their circulation means extra hemoglobin for carrying around oxygen. And the increased amount of blood means extra room for all those red blood cells.

If there is more fluid and red blood cells in the circulation, however, more blood vessels are needed to contain all that blood. One of the ways this shows up is by an increased density of capillaries in a mountain dweller's lungs—ideal for capturing extra oxygen with each breath. They also have larger, more powerful hearts than you would predict for someone of their body size. In addition, people and animals that live in the mountains even have differences at the cellular level. Their cells contain more of the energy-producing units called mitochondria, and each mitochondrion contains more of the enzymes needed to produce energy.

Thus, a person who is born and raised at a high altitude has remarkably different cardiovascular and respiratory systems than people living at sea level. This is especially obvious when a valley dweller tries to climb a mountain and develops what is known as mountain sickness. Without the special adaptations of llamas or native mountain inhabitants, climbing a very high mountain quickly results in a decrease in blood levels of oxygen and eventual collapse (even death) if oxygen therapy is not begun or a return to lower altitudes is not immediately started. If the initial climb is gradual, however, the body can usually adjust to high altitude by stimulating erythropoietin production, hyperventilating to get more oxygen, and numerous other changes. But no matter how long people live at high altitudes, if they were born and raised at sea level, they will never reach a stage of adjustment to low air pressure equivalent to those who have spent their entire lives in the mountains. No one is quite sure why this is true, but it is clear that the changes in anatomy and physiology of lifelong mountain dwellers begins in infancy, possibly even in fetal life.

One of the serious drawbacks of mountain climbing is that the blood pressure within the lungs can rise to high levels when oxygen pressure is low. This stems from changes in the diameter of blood vessels in the

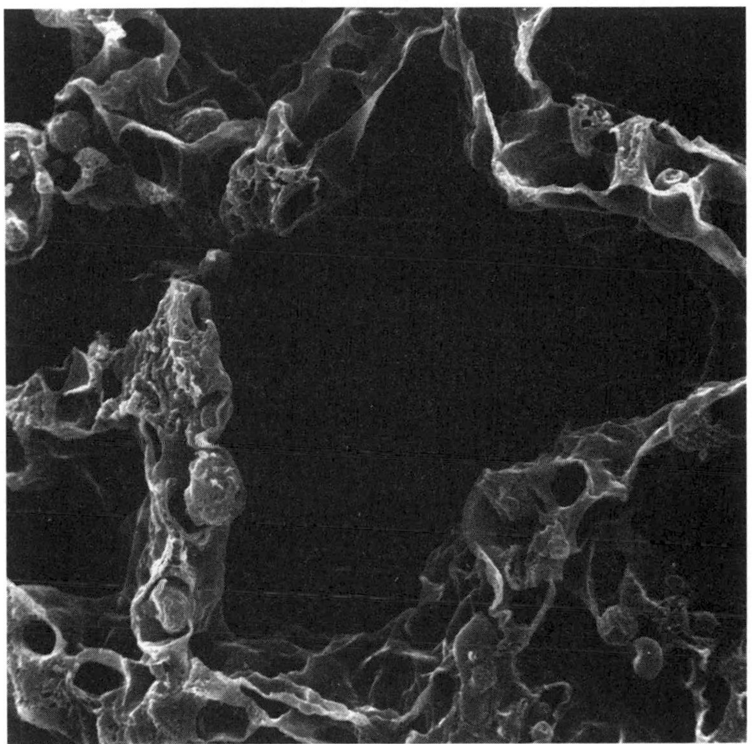

*Because it is in the alveoli, or air sacs, that we obtain oxygen and transfer it to the blood, it may seem illogical that the alveoli are comprised of the most delicate and fragile cells in the human body. This scanning electron micrograph illustrates the flimsiness and thinness of the alveolar cells. If the cells were not this thin, however, oxygen would have too great a distance across which to diffuse, and our blood oxygen levels would quickly drop to dangerous levels. Normally, these cells retain their shape and thinness throughout life, but in certain pathological conditions, the barrier to oxygen movement can be increased. In pulmonary edema (water in the lungs), the fluid layer between the alveoli and the nearby capillaries is expanded, resulting in a considerable diffusion barrier for oxygen, and even though the alveoli are normal, this fluid prevents oxygen from getting to the capillaries at a normal rate. Pulmonary edema is usually a sign of serious disease, such as that resulting from damage to the left ventricle of the heart.* *(Micrograph copyright © David M. Phillips/Visuals Unlimited. Used with permission.)*

lungs, and its consequences can be severe. When blood pressure rises in the lungs, some of the fluid that is normally contained in the capillaries leaks out into the spaces surrounding the tiny air sacs known as alveoli. This creates a barrier for the transfer of oxygen and carbon dioxide across the air sac membranes. This condition, known as pulmonary edema, makes it even harder to get oxygen into the blood and is what triggers many of the symptoms of mountain sickness. For unknown reasons, some people are far more susceptible to this problem than others, while still others seem to be almost totally resistant. Diuretics, which cause the body to lose extra water via the kidneys, should theoretically alleviate this problem by helping to lower blood pressure in the body, but the results of diuretic therapy in mountain climbers have been mixed.

Low oxygen pressure in the blood, resulting from pulmonary edema, can cause the blood vessels in the brain to open more, or dilate, thus bringing more blood to the brain each minute. Presumably, this is a natural response that keeps the amount of oxygen delivered to the brain more or less constant. The problem is when the brain vessels are dilated for a long time, you can get edema there, too. If edema in the lungs is dangerous, you can imagine how bad brain edema can be. This probably contributes to the mental confusion that often accompanies mountain sickness.

It is interesting to compare a diving mammal's coping mechanisms with those of a mountain dweller. In both cases, oxygen deprivation is the common thread. Whether oxygen is completely unavailable, as it is to us underwater, or is at a lower barometric pressure than normal, as at high altitude, a seal and a person respond in similar ways, by increasing blood volume, red blood cell number, oxygen-carrying ability, and other coping mechanisms. Without these remarkable abilities, all mammalian life would have been confined to a narrow range of terrestrial altitudes.

## The Bar-headed Goose and Mount Everest

Mammals are a poor cousin when compared to birds and one considers pulmonary performance at altitude. The best example is an improbable animal known as the bar-headed goose, and it is worth a closer look.

We are all familiar with the Herculean efforts of those few intrepid individuals who manage to climb to the summit of Mount Everest, more than 29,000 feet above sea level. Imagine, however, how the bar-headed goose must mock us as it effortlessly flies over the Himalayas each year during its annual migration. It needs no gas mask or gradual day-by-day ascents up short stretches of mountain; the goose takes flight and expends great amounts of energy through winds and bad weather to get to and from its wintering and nesting grounds. How does the goose manage such a feat?

Without going into great detail, the goose has a number of specializations, some of which are common to all birds. Although birds, like us, have lungs, their lungs look entirely different from our own. In fact, if it weren't that their lungs branch off of a trachea, you might not even recognize them as lungs at all. More than that, connected at random spots along the length of the lungs are air-filled sacs. People used to think that these air sacs were like balloons that gave a bird "lift" and helped keep it flying, but this is untrue. Adding air-filled sacs to an object doesn't make it lighter. You can only make an animal lighter if you substitute an air-filled space for something that was heavier than air. For example, many birds have hollow bones in parts of their bodies, in which the heavy bone marrow has been replaced by air. That does make them lighter and easier to stay aloft, but the lung air sacs of birds are something entirely different, because they actually permit birds to get oxygen even when exhaling.

The air sacs act as reservoirs that squeeze air through the lungs, much like a bellows. When a bird inhales, some of the fresh air goes into its lungs, and the rest goes into some of the sacs. Only the oxygen in the lungs can actually reach the bloodstream. But here's where the superiority of the bird respiratory tract comes in. What happens when we exhale? We get rid of stale air and prepare to take in some more fresh, oxygen-rich air, but while we're exhaling, no new oxygen is entering the blood vessels in the lungs. Thus, about half of our in-out breathing pattern doesn't contribute to getting additional oxygen. In the bird, on the other hand, even when it exhales it gets fresh oxygen. As the bird exhales, it not only gets rid of stale air from the lungs, but it squeezes fresh air out of the sacs and over its lungs. All air travels in one direction through a bird lung; it's a one-way flow-through system. Fresh air comes in through the trachea and enters the bottom of the lungs first. The rest of the air goes into the sacs. During exhaling, fresh air moves through the middle and top portions of the lung and out the trachea again. Meanwhile, the air sacs connected to the bottom of the trachea squeeze some more fresh air into the bottom of the lungs. So a bird can actually get fresh air both when inhaling and exhaling. This is far superior to our own system and is a major reason why birds can work so hard even where the air is thin.

Remarkably, fish have a similar type of one-way flow-through system in their gills but without the sacs. They, too, can get oxygen into their blood even when "exhaling," although the concepts of inhaling and exhaling are a little hard to define for a fish. So, while we humans are used to thinking of ourselves as the ultimate achievement of evolution, fish and birds put us to shame when it comes to one of the most basic needs of all animals—getting enough oxygen to survive.

Now that we know some of the problems faced by man and other animals in obtaining oxygen and eliminating carbon dioxide, let's see how these two gases are carried from place to place within the body, along with all the other nutrients we need to survive.

CHAPTER SIX

# Life Under Pressure

*The heart of animals is the foundation of their life, the sovereign of everything within them, the sun of their microcosm, that upon which all growth depends, from which all power proceeds. ... The King, in like manner, is the foundation of his Kingdom, the sun of the world around him, the heart of the republic, the fountain whence all power, all grace doth flow.*

—THE OPENING LINES OF WILLIAM HARVEY'S TREATISE ON THE CIRCULATION, *EXERCITATIO ANATOMICA DE MOTU CORDIS ET SANGUINIS IN ANIMALIBUS* (1628)

The great physiologist William Harvey—physician to two kings of England and a politically savvy man—was not above incorporating a little flattery into his most famous scientific work. It may seem hard for us to believe now, but as late as the sixteenth century, the role of the heart in the circulatory system was still uncertain, partly because it had yet to be realized that such a thing as a circulation even existed. There had been scattered reports before then that blood was transported from place to place—under pressure—in a vaguely predictable sort of way. But for 1,500 years most scientists and physicians were under the spell of the Galenists, who believed that the blood was produced from food in the liver, then pumped to the heart.

Harvey's major contribution was to demonstrate irrefutably that the Galenist view (named after the second-century Greek physician who first advanced the idea;

*It took 1,500 years for scientists and physicians to escape the traditional view of the circulation first elaborated by the Greek physician Galen. Although in many respects Galen is still considered one of the founders of modern anatomy and physiology, some of his conclusions regarding the heart and blood vessels were, by today's standards, fanciful. For instance, Galen believed that the liver played a significant role in pumping blood to the heart and that food was directly converted into blood by the liver. (Courtesy of the Francis Countway Medical Library.)*

*The British physician William Harvey was the first to clearly document that a fixed quantity of blood circulates over and over again within closed vessels, and that the heart was the only source of power to pump the blood around. In doing so, Harvey laid the groundwork for modern cardiovascular medicine. (Courtesy of the Francis Countway Medical Library.)*

see Chapter 1) of the way in which blood was formed and pumped through the body was incorrect.[1] Instead, Harvey proved that blood was self-contained in vessels and recirculated over and over again throughout a person's lifetime. This laid the groundwork for all we know today about cardiovascular medicine.

## The Origins of the Circulation

What was it about Harvey's idea of the circulation that was so important? We demonstrated in previous chapters the ability of the body to regulate its energy content through the processes of eating and breathing. Both of these functions would be meaningless, however, without a way to transport the fuels and oxygen from their ports of entry (that is, the intestines and the lungs) to the places they're needed (essentially, everywhere). This is a truly heroic task, when one considers the number of living cells in an animal our size (between 100 trillion and 1 quadrillion) and how distant many cells are from the intestines and lungs. For example, the muscle cells of our big toes are roughly 4 feet away from the lungs, an impossibly long distance for oxygen to travel without a circulatory mechanism.

We see, therefore, that by understanding the features of the circulation, we can better understand how nutrient flow to all our cells is regulated. But how did circulatory systems first evolve, and what is it about our own system that is so special? The answer depends on how a circulatory system is defined. Jellyfish, for example, have no blood vessels or heart as we know them, but they do have a circulation. The "bell" of a jellyfish is composed of muscles and nerves. In order to swim, jellyfish contract the muscles of the bell and propel themselves along, instead of relying on ocean currents to drift aimlessly (although they do that, too, part of the time). When the bell relaxes, it expands and draws water up into a network of canals. When the bell contracts, seawater is squeezed back out of the canals. The canals percolate through the entire muscular bell, so that each cell in the jellyfish comes into contact with seawater. Seawater can be considered the "blood" of the jellyfish, because it contains oxygen and some nutrients, and the canals can be thought of as their blood vessels. The rhythmic beating of the bell—so obvious when you see a jellyfish swimming in the water—is analogous to the beating of our hearts.[2] So, when a jellyfish needs to swim faster, the bell contracts more frequently, which automatically ensures that oxygen will be circulated through the bell at a faster rate. This is exactly what happens to us when we increase our own activity level. As we exercise, our hearts beat faster to increase the rate at which oxygen and nutrients will be delivered to our cells.

Although a jellyfish's primitive means of circulating nutrients is barely comparable to our own, true circulatory systems that are similar are found in some invertebrates, and can even be quite sophisticated. The cephalopods (squid and octopus) have extremely complex circulations that enable them to be more active than most invertebrates. But it's in the vertebrates, like ourselves, that the circulation reaches its apex.

## Pumps and Vessels

Just what is needed to make a complete circulatory system? In large animals like ourselves—and, in fact, in anything much larger than a worm—a pump is needed to move blood around. To force blood against gravity up to the top of our head is a major undertaking. The vertebrate heart—although it varies in shape from fish to mammals—does essentially the same thing in all animals. In mammals, the heart is composed of four chambers, two upper ones (atria) and two lower ones (ventricles). The atria are filling vessels that collect blood and direct it into the ventricles. The ventricles are much more massive than the atria, and are responsible for pumping the blood throughout the body. Strong, tireless muscles comprise the walls of the ventricle and permit it to contract, or beat, with a regular rate and with great force. That it can do this billions of times without rest and without getting fatigued is one of the great wonders of nature. Imagine squeezing a rubber ball in your hand once each second, night and day, without fail, for 70 years. That's how hard the heart works.

As blood is propelled from the heart into the arteries, it generates a wave of pressure, just like the pressure generated in a garden hose when the faucet is turned on. In animals whose heads are higher than their hearts—like us—that pressure must be great enough to drive blood all the way up to our brains. As discussed earlier, the blood pressure of a giraffe is higher than our own, and its heart is truly mammoth (about the size of a soccer ball). In fact, were the neck of a giraffe to get much longer, the circulatory system might not be able to withstand the pressure needed to pump blood that high. We can only speculate on blood pressures in extinct dinosaurs, some of which had necks up to 30 feet long!

With occasional exceptions, the size of the heart increases in direct proportion to the size of the animal. Remarkably, in almost every case the heart makes up about one-half of 1 percent of an animal's body weight. And the bigger the heart, the stronger it is. Even within our own species, heart size can be increased or decreased, depending on how active we are, just like the muscles in our arms and legs. A very active person who keeps in good cardiovascular shape has a heart that is bigger than normal. Conversely, if that person were to stop exercising and lead a more sedentary life, the heart muscles would shrink.

*The advantage of a long neck for giraffes is easily seen in this photograph, taken on the plains of Kenya. The disadvantages are not as obvious. To pump blood from the heart to the head, against gravity, requires an enormously powerful heart and very high blood pressure. By the time the blood reaches the brain, the pressure is much lower than that near the arteries as they first leave the heart. Even in a human being, with a head only about a foot higher than the heart, the blood pressure reaching the brain is much lower than that reaching, for example, the legs. This prevents the brain from swelling due to excess pressure. Nevertheless, the giraffe heart must work considerably harder than that of comparably sized animals with normal-length necks. (Photo courtesy of Dr. Charles K. Levy.)*

It is worth noting that there is a pathological way in which the heart can enlarge because of excessive strain. In people with high blood pressure (hypertension), the heart must work harder to pump blood against that wall of pressure. Thus, the heart enlarges. This situation is known as ventricular hypertrophy, because it is the lower, or pumping, chambers of the heart (the ventricles) that enlarge (hypertrophy). Eventually, the strain on the heart can become too much to bear, and a heart attack may ensue. Nowadays, doctors can determine how thick the heart muscle has become using an echocardiogram, a noninvasive procedure. Using this information, they can roughly deduce how long the hypertension has been present, which provides a clue to the severity of the disease and aids the doctor in developing a treatment plan for the patient.

A curious problem arose some years ago when doctors tried to use the technique of measuring ventricular thickness to analyze the cardiovascular fitness of elite athletes. Most athletes have hearts that are considerably enlarged and very powerful. Nonetheless, even elite athletes are not immune to developing high blood pressure. The problem for physicians was how to diagnose the duration and severity of hypertension in people whose hearts were already enlarged due to chronic aerobic exercise. One possible solution to this paradox was to collect sufficient data from a large number of healthy athletes in various sports, so that a baseline range of ventricular thickness could be established with confidence. A study of this sort was conducted in 1991 by the Italian National Olympic Committee together with researchers in Rome, using the members of the Italian Olympic team as subjects. Their findings established guidelines for discerning between exercise-induced ventricular hypertrophy and that induced by chronic hypertension.[3]

As William Harvey suggested, the heart is the structure around which the circulation is built. What other components are required for a complete circulation to work? First of all, vessels of some kind are needed to carry the blood on its journey. The thick, strong, tubelike vessels that carry blood away from the heart are the arteries. Were an artery to rupture, blood would spurt out under high pressure, and with each beat of the heart, a new pressure wave would be generated, forcing more blood out of the artery. Without prompt medical attention, a ruptured artery could be rapidly fatal. Not surprisingly, therefore, animals like ourselves have evolved arteries that are normally buried deep below the skin and not susceptible to the minor cuts and bruises we commonly encounter.

The walls of arteries are too thick to allow the nutrients and oxygen needed by cells to diffuse across them. Oxygen, sugar, hormones, vitamins, and everything else carried by the blood are trapped within the arteries and can't penetrate the arterial wall. But as arteries branch off into smaller and smaller arteries, called arterioles, they eventually reach a point where they can't get any smaller. These tiny vessels are called capillaries, and their diameter is roughly the width of a single blood cell (about 3 ten-thousandths of an inch). Capillaries, unlike arteries, are leaky and thin-walled; in fact, some capillaries are dotted with microscopic holes in their walls that allow nutrients and other molecules to escape. There are capillaries in close proximity to virtually every cell in the body, and it is from the capillaries that the oxygen and other molecules leave the bloodstream and percolate into the fluid-filled spaces around cells. Once there, they can be captured by cells and used to drive the chemical reactions and other activities that are needed for survival.

As oxygen and the other components leak out of the capillaries, they draw fluid from the blood with them. The fluid that leaves the capillaries and bathes the cells must be recaptured by the circulation, or it will result in a very uncomfortable situation. As we saw in Chapter 5, when excess fluid builds up in the spaces between blood vessels and cells, it's called edema. Many of us—especially when not being very active—have experienced the sensation of our shoes feeling tighter at the end of the day. This is because gravity is pulling on our circulation and causing excess fluid to leak out of capillaries in our legs and feet. This problem is easily remedied, however. By flexing your leg muscles occasionally throughout the day, you help squeeze the excess fluid back into the capillaries and veins where it belongs, where it can then be carried back to the heart, which takes us to the next stage of the circulation.

Normally, the capillaries recapture the fluid that leaked out of them and dump it into the next set of vessels, the veins. Anything missed by the veins is carted off by the lymph vessels. Veins are floppy, distensible structures that can fill up with enormous quantities of blood. Because of their compliant nature, there is very little pressure in the veins, unlike the situation in the arteries. In addition, the kind of pressure that was in the arteries dissipates as fluid leaks in and out of capillaries. This is why, when we cut a vein, blood oozes out slowly, rather than pulsing out in great bursts.

The veins bring blood back to the right side of the heart through two large vessels. Veins above the heart pool into the superior vena cava, and veins below the heart pool into the inferior vena cava. These two large vessels collect into what becomes the right atrium. The veins below the heart have one-way valves within them—directed upward. When we contract a leg muscle, the veins are compressed and blood is pushed upward. The valves prevent backsliding of blood.

From the right atrium blood flows into the right ventricle and is pumped out again. But what is the destination of blood coming from the right ventricle? Having passed through the body and given up much of its oxygen, this blood needs another burst of fresh oxygen before it can be of further use, and it needs to unload the carbon dioxide that it picked up from the body's cells. Therefore, the blood from the right ventricle is routed directly to the lungs and nowhere else. Here, it picks up oxygen and drops off carbon dioxide. The rejuvenated blood flows out of the lungs and back to the other side of the heart, into the left atrium. Finally, blood makes its way to the left ventricle and is pumped out again to the rest of the entire body. Since the left side of the heart has a bigger job to do, it is much larger than the right side.

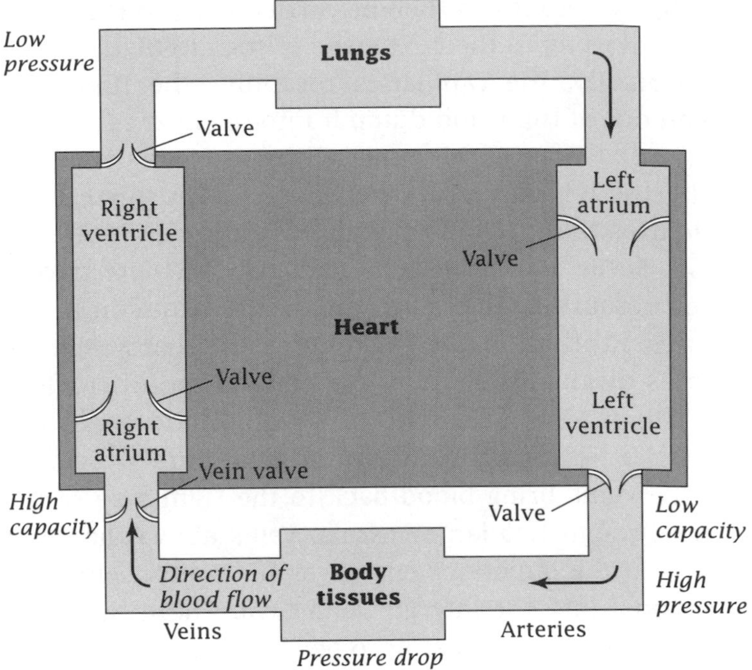


*Although our heart has four chambers, all seemingly connected and in one place, from a functional point of view, the two sides of the heart can be considered as two separate hearts. In fact, during embryological development, the heart begins as two separate tubes that fuse together, leaving a thick septum, which divides them thereafter. The only reason the two sides need to be together is to allow a single pacemaker to drive the rhythm of both sides simultaneously, ensuring that the left and right ventricles pump blood at the same moment. In this sketch, we see the route blood takes as it leaves the left ventricle, travels through the arteries under high pressure, percolates through the capillaries where oxygen, carbon dioxide, and nutrients are exchanged with the cells of the body, then returns via veins under low pressure to the right heart. From there, blood is pumped by the thinner-walled right ventricle under low pressure to the lungs, where oxygen is picked up and carbon dioxide is lost. Blood then flows directly back to the left heart. Thus, functionally speaking, the two sides of the heart form what's known as a series circuit with each other, with no crossover or mixing of the two sides of the circulation.*

And the artery that receives the blood from the left ventricle, the aorta, is the biggest one in the body.

This is the situation that Harvey knew had to exist, even if he couldn't actually see any capillaries. Blood is recirculated again and again, picking up oxygen in the capillaries in the lungs and dropping it off in the capillaries in the rest of the body. Does this mean that the blood in our system is unchanged from birth to death? Certainly not. Both the fluid component of blood (plasma) and the cellular component (red and white blood cells) are regenerated over and over again. For instance, if we become dehydrated, the amount of water in our plasma drops. When we take a drink of water, the water enters the blood and becomes part of the plasma. Likewise, blood cells are constantly dying and being replaced.

## Generating Pressure

To make a circulatory system work properly, you need a steady source of pressure. We've seen that the heart is a major source of this pressure, but what keeps the heart going? And what regulates how hard it pumps? The answers to these questions are virtually the same in every vertebrate, including ourselves. Within a region of the right atrium is a specialized collection of heart muscle cells known as the sinoatrial node or, colloquially, the heart's pacemaker. These cells drive the rhythm of the heart to produce the characteristic heart rate of a given species. In general, the smaller the animal, the higher its heart rate, even within a single species. Children have higher heart rates than adults on average, and women normally have slightly higher rates than men.

Every individual cell of the heart could act as a pacemaker if allowed to. If a dissected heart of a frog, for example, is treated with enzymes that break the heart up into its millions of individual living cells, an astounding thing happens. When grown in a petri dish, each heart muscle cell beats with its own intrinsic rate, as if the heart was really made up of millions of microscopic hearts all sewn together. It turns out that of all those cells, a few will beat faster than all the others; these are the pacemaker cells. Once they beat, they set off a wave of electrical excitation that causes all the other miniature "hearts" to beat, too. It is when these cells are damaged that an artificial pacemaker must be implanted under the skin and connected to the heart, or else the beating of the heart would become erratic.

The pace of the natural pacemakers can be changed, too, when needed. During exercise, for example, neural signals from the brain travel to the heart and speed up the pacemakers; this makes the heart beat faster. Our heart rate can reach about 180–200 beats per minute, or three beats per second. Any faster than that and the heart wouldn't have sufficient time to fill with blood between beats, and we'd quickly lose consciousness. But very tiny animals, such as hummingbirds, have heart rates that can reach 1,300 beats per minute. That means that each beat of the heart, including the refilling of the heart with blood, lasts for only about one-twentieth of a second!

Even emotional distress can alter our heart rate. Excitement, anticipation, joy, and so on all speed up our heart. In some people, the opposite emotions of grief, sorrow, fear, and shock can slow the heart down. This condition, known as vagovagal syncope, is an abnormal one in which an individual faints due to the light-headedness brought about by a slowed-down heart. Fainting occurs because the brain fails to get sufficient oxygen and fuel. Fortunately, fainting is exactly what is needed, since the head and the heart would then be at the same level, making it easier for blood to reach the brain.

The same signals that speed up or slow down the heart can also increase or decrease the strength with which it beats. This, too, will contribute to blood pressure, as will certain substances we eat or use as medication. Caffeine and its chemical cousin theophylline (found in tea) both excite the heart, as do some over-the-counter diet pills. Nasal sprays taken to relieve congestion resulting from a cold or allergy may also speed up the heart and raise blood pressure. In fact, any cold medication that has the word "decongestant" on it may have some effect on elevating blood pressure. People with high blood pressure are sometimes advised to avoid such cold remedies.

As you might have guessed by now, blood pressure is not the same everywhere in the body. The pressure in our feet is much higher than that going to our head, because of gravity. This usually only becomes noticeable when our blood volume is lower than normal. Take, for example, the case of a person exercising on a hot day. All that perspiring creates a situation where fluid leaves the body in great amounts (a pitcher in a baseball game can lose 5 to 10 pounds during the course of a game on a hot summer day). Most of that fluid is water and can be replaced immediately by drinking. But anyone can become mildly dehydrated on a hot day, just from usual daily activities, and most people do not drink as much fluid each day as they should. The result is that we tend to be slightly dehydrated for much of our lives. If this is compounded by excessive sweating, we can become quite faint. In fact, almost everyone has experienced the light-headedness that follows when we jump to a standing position after lying or sitting down for some time. This happens more often in summer, for the reasons just stated. When we are dehydrated to begin with, it is harder for the heart to generate sufficient pressure to fight off the effects of gravity. Thus, when we stand up suddenly, our brain doesn't get as much blood as it needs for a few seconds, and we get dizzy. In normal people, there are internal defenses that quickly compensate for gravity and prevent us from actually fainting, but in individuals whose blood pressure tends to be on the low side anyway, fainting occasionally occurs. The situation is even worse in some disease states in which the nervous system is affected and is known as orthostatic hypotension.[4]

## Measuring Blood Pressure

When we are given a physical exam at the doctor's office, blood pressure recordings are usually taken. The doctor gives you two numbers, such as "120 over 80 (120/80)." These numbers refer to the ability of your blood to raise a column of mercury under pressure. If your blood pressure is sufficient to raise the mercury 120 millimeters, we say your pressure is 120 millimeters mercury. The top number is the systolic pressure, which occurs when the heart muscles contract and pump out blood. The second, lower number, or diastolic pressure, is reached while the heart relaxes between beats. You might think that the higher of the two numbers would be the most important in determining your overall cardiovascular health, but it's often the diastolic number that matters most. That's because the relaxation phase of the heart lasts for twice as long as the pumping phase. So, if the lower number is elevated, it can be twice as significant than when the higher number is elevated. The lower end of normal pressures—found more commonly in women than men—is around 100/60, while the upper end of normal is around 130/85. The pressure tends to increase a bit as we age and is much lower in young children and infants. According to guidelines published by the National Institutes of Health, high blood pressure can be divided into mild (140/90–159/99), moderate (160/100–179/109), severe (180/110–209/119), and very severe (210/120 and higher). But don't be fooled by the terms. Even mild hypertension can be lethal given enough time. In fact, cardiovascular diseases like hypertension account for roughly one-third of all deaths in the United States, far greater than deaths due to cancer, the number two killer.

If a doctor discovers that your blood pressure is high, he or she first tries to determine why. Unfortunately, in a very high percentage of cases, the cause of the elevated pressure cannot be determined. That makes it difficult to establish a good treatment plan, and often the only recourse is to prescribe medication. According to the U.S. Department of Health and Human Services, doctors write some 50 million prescriptions each year for medications to control heart and vessel diseases in the United States, more than for almost any other category of illness. Except in unusual cases, people with mild high blood pressure (say, 145/95) respond very well to these medications. A common class of drugs prescribed today is one that acts by inhibiting the body's ability to make excess amounts of the hormone angiotensin II (see note 4), or AII. One of AII's normal functions is to prevent blood pressure from falling below normal levels, but in some people with hypertension this hormone apparently works too well, resulting in a pressure that is above normal.

On the other hand, there are many people in whom it is possible to deduce the cause of their hypertension. For many years now, a diet high in salt has been believed to contribute to hypertension, possibly by elevating the total fluid level in the bloodstream. The jury is still out on this theory, because some people appear to respond favorably to low-salt diets, while others do not. Still others mistakenly reduce their salt content so much that they develop additional problems. Too little salt can be almost as bad as too much.

Perhaps the easiest way for many people to reduce their high blood pressure is to exercise and lose weight. My own blood pressure dropped from 145/95 to 125/80 when I dropped a few pounds and improved my diet. The motivation was simple: Either start taking medication for the rest of my life, or get on the treadmill and eat a healthier diet. No one had to tell me twice which was the better alternative.

What is so awful about high blood pressure? The only real symptom that someone might be aware of if their pressure were moderately higher than normal would be a tendency to develop headaches. Over the years, however, the high pressure creates such a strain on the heart that it wears it down. High blood pressure is also one of the three major factors contributing to the development of atherosclerosis—smoking and cholesterol are the others. High blood pressure can also lead to stroke and weakening of artery walls.

## Shock

While we can die from high blood pressure, we can also die from low blood pressure! I've already mentioned that low blood pressure predisposes a person to orthostatic hypotension, but what if pressure were to suddenly become really, really low? An example of such a situation is someone who has lost a great deal of blood due to a severe injury. As blood leaks out of the circulation, the pressure in the circulatory system rapidly plummets. Unless something is done to prevent the leakage, the individual may quickly slip into shock, or circulatory collapse. Shock can be defined as a state in which blood pressure is too low to provide oxygen and nutrients to the cells of the body at a level needed to sustain their function. All mammals enter shock under these conditions, some sooner than others, and in each case (including our own), the body responds in essentially the same way to prevent shock from progressing to a lethal stage.

Like the chemoreceptors that tell the brain when oxygen is too low, there are special blood pressure detector cells, called baroreceptors, within the walls of the aorta and carotid arteries. Actually, the usual role of the baroreceptors is to prevent blood pressure from rising above normal. Thus, when an artery is under high pressure, its walls become distended. This, in turn, stretches the baroreceptors, which then become activated. Signals from the baroreceptors are sent to the brain. The brain interprets the message (Warning! Pressure is too high) and sends signals of its own to the heart, causing it to slow down. As the heart becomes less active, blood pressure starts to fall back toward normal.

Conversely, when pressure falls below normal because of a loss of blood, the walls of the arteries are less stretched than normal, which means that the baroreceptors are no longer stretched and therefore do not become activated. Thus, input from the baroreceptors to the brain is decreased; the brain stops trying to slow the heart, and the heart speeds up. This helps restore pressure.

But simply speeding up the heart is not going to compensate for the fact that blood was lost from the circulation. The fluid that was lost must still be replaced or the pressure will never return to normal, no matter how hard the heart works.

To do this, some fluid shifts out of the cells and into the bloodstream. In addition, a strong sensation of thirst is activated, triggering the desire to drink, and drinking, of course, helps restore blood volume. Many people that are clinically in shock remain fully conscious. In fact, it is often better for the patient to be conscious, because the anxiety associated with the injury keeps the person breathing at a high rate and keeps their metabolism going. As shock progresses, consciousness is lost, and the situation becomes much more grave. Eventually, shock can become irreversible, at which point no treatment of any kind can restore pressure to normal and recovery is impossible.

As long as shock has not reached that final stage, however, fluid can be restored with an intravenous drip.

Injections of natural hormones are also given to further stimulate the heart. These hormones, such as epinephrine, also cause blood vessels to narrow, which also helps raise pressure. Other hormones, like cortisol (hydrocortisone), potentiate the effects of epinephrine and help protect the body against the stress of the injury.

A person suffering from this type of shock may feel cold, because the blood going to the skin has been shunted elsewhere. All mammals, including ourselves, have evolved the ability to decrease the amount of blood flowing to less vital regions and increase it to important regions when pressure is low. Thus, the skin, intestines, and kidneys are deprived of blood for a time, while much of the rest of the blood is rerouted to the heart and brain.

Over the days and weeks following a large hemorrhage, the liver produces high amounts of proteins that enter the circulation and help draw additional fluid out of the cells and into the plasma. The bone marrow is stimulated to produce more red blood cells, which contain the hemoglobin needed to carry oxygen. Complete recovery from a massive hemorrhage in the absence of a blood transfusion can take weeks.

Hemorrhage is not the only event that can cause circulatory shock, although it's certainly the one with which we are most familiar. There are ways in which shock can develop even without the loss of a single drop of blood. One of these is the condition known as anaphylactic shock, which results from a massive allergic response to certain antigens, like bee stings or shellfish (actually, bits of the shell itself). In people that are prone to allergic reactions of this type, the body releases large amounts of chemicals like histamine that make blood vessels wider and leakier, so wide and leaky, in fact, that they lose the ability to hold blood under pressure, as well as the ability to redirect blood to vital areas. Blood that should be flowing to the heart and brain now flows elsewhere, such as to the skin. That's why in this type of shock the skin feels warm and not cold.

Anaphylactic shock is especially nasty, because histamine also causes bronchoconstriction, especially in asthmatics. Many people with allergies also have asthma, so when histamine is released into the blood, the muscles around the airway tubes (bronchioles) start to contract, closing off the airways. This syndrome is easily treated with antihistamines and epinephrine. The problem comes with the first attack. If not prepared for it, a person stung by a bee could die within minutes if anaphylaxis were to develop. Nowadays, most people who have had an anaphylactic experience carry an injectable form of epinephrine and antihistamines that can be self-administered in the event of a new attack.

Perhaps the worst type of circulatory shock of all, however, is congested shock. What gets congested are the lungs, and the cause of the congestion is failure of the heart. If the heart muscles do not get enough oxygen, they start dying. This happens more frequently in the left ventricle than any other place, because this is the hardest-working part of the heart. If part of the left ventricle dies, the remaining tissue can't beat as strongly as before. The blood that enters the heart from the lungs won't be pumped out through the aorta as efficiently as before. The blood backs up because it has nowhere to go, and this creates pressure in the lungs. To understand this, think of a garden hose. Imagine it's turned on full blast, and someone pokes a small hole in it halfway between the faucet and the nozzle; water will gush up through the hole like a miniature geyser. If you were to put a clamp around the nozzle end of the hose—past the hole—while leaving the faucet on, what would happen? The water would back up, and the geyser would shoot up higher under greater pressure. This is exactly what happens in our body after a left ventricular heart attack. Only now, the geyser is in our lungs! This is the last place you want fluid to leak into.

As fluid accumulates in the lungs, it makes it harder for oxygen to move from the alveoli into the blood vessels. The fluid creates a barrier to oxygen movement. Not surprisingly, therefore, one of the symptoms of a left heart attack is shortness of breath. In fact, many people have a mild left heart attack without ever realizing that's what's happening to them. Later, they go to the doctor complaining of shortness of breath. Climbing stairs, taking a walk, and other trivial activities leave them breathless. This is an important clue for a doctor—together with such other considerations as age and fitness—and suggests a problem with lung congestion that may be secondary to heart disease.

If the heart attack is severe enough—and the congestion great enough—shock ensues. Only about 15 to 20 percent of people who develop congested shock manage to survive. Remember, one of the main ways the human body restores pressure during shock is by increasing the activity of the heart, and this mechanism obviously can't work as well if the heart is damaged.

With all the things that can go wrong with us, from daily aggravations to dehydration, plaque buildup in our arteries, sickness, injuries, poor diets, and so on, it's a wonder that we don't all suffer from hyper- or hypotension. It's truly a testament to the power of homeostasis and the way in which the human body has been designed to counteract changes imposed upon it.

Besides shuttling food and oxygen around the body, however, there is another important function of the bloodstream. We and many other animals use our blood to either conserve heat or, on hot days, to cool down. In fact, the way in which we do this has provided a clue as to whether or not dinosaurs were warm-blooded animals like ourselves. We'll find out how blood affects body temperature in the next chapter.

CHAPTER SEVEN

# Bat Wings and Elephant Ears: Keeping Cool

*Our ideas must be as broad as nature if they are to interpret nature.*

—SIR ARTHUR CONAN DOYLE, ENGLISH AUTHOR, IN *A STUDY IN SCARLET* (1888)

It's hard to imagine that bat wings and elephant ears could have much in common. After all, wings are for flying and ears are for hearing, and the entire body surface of a small bat isn't even one-tenth the size of an elephant's ear. But in fact these structures are related in a simple, but important, way that will determine an essential function for each—both are very thin for their size. Some large bats, for example, have a wing span of up to 4 feet (now that's frightening!), but the wings are still thin enough to be nearly transparent. This means that these structures have a good deal of surface area relative to their size. Recalling what we learned about surface area–body mass relationships in Chapter 2 (remember why babies need to be kept warm?), it's easy to predict that such structures are well suited to act as heat radiators. Indeed, in addition to their more obvious functions, bat wings and elephant ears are important in maintaining body temperature. In this way, they are also similar to a rabbit's ears, a cow's udder, and our own skin. To appreciate how surface area contributes to the ability to radiate (or dissipate) heat, it's worth understanding why radiators are needed in the first place.

Of the countless ways to categorize animals, one of the simplest and most common is to assign them into two major groups, namely warm-blooded or cold-blooded. The defining feature of warm-bloodedness is the ability of an animal to keep its body temperature fairly stable no matter how cold or hot it is outside. This is harder for animals like ourselves, because we don't have an insulating layer of thick fur or feathers or blubber like seals. Nonetheless, we and other warm-blooded animals do a very good job of maintaining our body temperature within narrow limits. Most warm-blooded animals—birds and mammals—have a body temperature between 95 and 105 degrees Fahrenheit, or about 35 to 40 degrees Celsius.[1]

Cold-blooded animals, on the other hand, can't produce their own internal heat the way we can. In order to warm up, therefore, reptiles, amphibians, and fish need to be in a warm environment. That's why a reptile likes to bask in the sun. The heat from the sun penetrates its body and raises its temperature. In fact, on a really hot day a reptile's body temperature can even exceed our own. The difference between us and cold-blooded animals, however, is that once the sun goes down, a reptile's temperature falls, whereas our body temperature remains at about 98.6 degrees.

Remarkably, some amphibians, such as the wood frog, and even some reptiles, can tolerate freezing and thawing, but such species are rare in nature. A more common strategy is that employed by the Antarctic icefish. This animal, which is cold-blooded, like all fish, inhabits some of the coldest waters imaginable but doesn't freeze. Circulating within its bloodstream are highly complex antifreeze molecules that prevent its blood

*This crocodile is not making a threatening display nor is it yawning. By opening its mouth and facing the sun, this cold-blooded reptile absorbs considerable heat, which it uses to power its otherwise sluggish metabolism. The mucous membranes that line its mouth are excellent heat absorbers, and some crocs will maintain this position for hours on end. (Photo courtesy of Dr. Charles K. Levy.)*

from turning into ice. There are a variety of these antifreeze molecules in nature, but they all do essentially the same things—lower the freezing point of blood and stop growing ice crystals from getting larger. The former property is related to the property of water whereby any substance, whether it is a quantity of table salt, sugar, protein, or anything, is dissolved in water, it will no longer freeze at 32 degrees F (0 degrees C) but will stay liquid until some new, lower temperature is reached. This is what happens when you add antifreeze to your car's radiator. The antifreeze lowers the temperature needed to freeze the water in the radiator, allowing it to remain a liquid even on very cold winter days.

Scientists have even perfected ways of freezing down suspensions of individual cells isolated from different tissues of the body. By adding a solute like glycerol to the cell suspension solution before freezing, the cells themselves are protected from ice crystal formation. They can be stored in a freezer for indefinite periods of time and taken out and thawed when needed. Of course, this does not mean that it is possible to freeze and thaw a complete person. We're still a long way from suspending a complex organ such as a heart or brain, let alone an entire human body.

## The Pros and Cons of Being a Reptile

It's easy to imagine both the advantages and disadvantages to being either warm- or cold-blooded. Let's discuss the cold-blooded animals first. Why do they need to warm up in the first place? Because nearly every chemical reaction in nature runs more efficiently at higher temperatures. Such reactions include those that cause muscles to contract, the heart to beat, the lungs to expand, nerve cells to fire, and fuels to be converted into energy. The energy compound that is produced by burning fuel and that drives all the aforementioned reactions is ATP. Without a steady source of ATP, no animal could survive, and if ATP is present at lower than normal amounts, many chemical reactions slow down or stop altogether. Even the production of ATP itself speeds up at higher temperatures. This puts a cold-blooded animal at a real disadvantage. Without adequate heat, it cannot make enough ATP, which, in turn, is needed to drive other reactions.

Thus, a reptile must spend considerable time soaking up the sun in order to get its metabolism revved up. This behavior brings with it certain perils, because the animal is exposed to predators during this time, and a basking animal hasn't built up sufficient stores of ATP to be active enough to elude a hungry predator for long if one were to appear.

Because cold-blooded animals must take time to warm their bodies each day, it also means they have less time to devote to other activities, such as defending territory, finding a mate, and finding food. On the other hand, if part of your life is spent at a lower temperature, you need less overall energy, because low temperature means slower chemical reactions. That's exactly why snakes and other reptiles sometimes can go for weeks without food.

Fish follow a similar pattern. With some special exceptions, their body temperature matches that of the surrounding water. Tropical fish, therefore, have higher body temperatures than fish swimming in Arctic waters. Not surprisingly, this means that many tropical fish are more active for longer periods of time than their cold-water relatives.

One well-known exception is the tuna. These large fish are like sharks in that they must swim constantly in order to breathe. To propel those large bodies around 24 hours a day requires some pretty serious musculature. To keep those muscles active enough to do all that work, they need to have a high rate of metabolism, which means lots of energy consumption and oxygen usage. To drive all that metabolism, however, heat is required. But because the tuna has a high rate of breathing, and cold water is constantly flowing over its gills, sapping most of the animal's heat with it, how does it keep its muscles warm?

The answer is something called a countercurrent heat exchanger. As blood enters the large swimming muscles of the tuna, it is warmed a bit by the heat generated during muscle use. This slightly warmed blood leaves the muscles in veins, which are arranged in tuna in such a way as to pass extremely close to the arteries that are entering the muscles with fresh blood. Thus, you have two tubes running in opposite directions, one into the muscle and one out. Heat transfers very rapidly across the surfaces of blood vessels, so, as the warm blood leaves the muscles through the veins, its heat transfers across the veins to the cooler arteries. This warms the blood in the arteries, which means that the blood entering the exercising muscle is warmer than it otherwise would be. Since the rate of chemical reactions, including those that drive muscle function, are increased at higher temperatures, the warmer blood of the arteries results in warmer, more active muscle cells.

When the muscle cells warm up, they work harder, which generates yet more heat, which makes the venous blood warmer, which transfers more heat to the arteries and makes them even warmer, which makes the muscles yet warmer, and so on! The final result is that although a tuna is a cold-blooded animal, it is often called "warm-bodied," because it has this fantastic mechanism to keep its swimming muscles (which make up most of the weight of a tuna) as warm as the muscles of a warm-blooded animal like ourselves. Interestingly enough, a similar heating mechanism is found in certain insects, such as the honeybee.

Of course, a cold-blooded animal has the reverse problem on a hot day. Under such conditions, its body temperature can reach extremes that could cause an internal meltdown. The best way to prevent this is simply to avoid the sun by remaining in hiding except at the cooler times of morning and night, which is exactly what many do.

## Warm-bloodedness: Life in the Fast Lane

We can see that being cold-blooded has some great disadvantages. Is the same true of warm-blooded animals? Humans have the advantage of being able to keep chemical reactions at full throttle, regardless of the outside temperature, which means they can be active at any time of the day or night, winter or summer. But is there a downside to being warm-blooded? One consequence of having thousands of energy-demanding chemical reactions occurring at high rates all the time is that energy stores in the body must be constantly replenished. And the only way to do that is by eating a great deal of food, every day. This may be easy for modern man, because we can walk to the kitchen whenever we're hungry, but for most birds and mammals, it means constantly hunting for food. The extreme case is the shrew, which, as we've said, is basically an eating machine.

Another big problem with being warm-blooded is the threat of overheating. As we all know, this happens easily if we exercise too strenuously on a hot day. Since our body temperature starts out at a high level compared to that of a cold-blooded animal, the temperature can rise to dangerous levels if we overexert ourselves, possibly resulting in what is commonly called heat stroke. In fact, if our temperature rises only a few degrees, we may become delirious. And while many people can tolerate fevers up to 104 or even 105 degrees, above that point—a mere 6 percent increase—adults start to become disoriented, risking permanent brain damage or even death.

This delicate temperature balance is not just a human concern. Whether you are a person, a goat, a rhinoceros, a bat, or a bird, if your body temperature goes up much more than a few degrees, you're in real trouble. At those temperatures, chemical reactions may proceed too fast, special proteins called enzymes become nonfunctional, and the ability of brain cells to function is compromised.

One way to solve this problem is to create built-in radiators in warm-blooded animals. Remember the elephant and the mouse and the issue of surface area versus body mass in Chapter 2. An elephant has much more area over its surface than a mouse, of course, but compared to its body mass a mouse's surface area is relatively greater than the elephant's. This means that large animals tend to retain heat better because they have proportionately less area over which the heat can escape.

To protect against overheating (hyperthermia), many birds and mammals have adapted certain parts of their bodies to act as radiators in dissipating body heat. For example, if the wing of a bat is stretched, it becomes readily apparent that the membranes that form the wing are extremely thin. Thus, they have lots of surface area but only the tiniest weight. Likewise, an elephant's ears are enormous as ears go, but they, too, are fairly flat with lots of surface area. Heat travels outward on these structures and quickly escapes into the surrounding air. The same can be said for human skin, a rabbit's ears, a goat's horns, and so forth.

Although we have discovered the device nature has created to lose excess heat, there is still the question of how the heat is conveyed from the various parts of the body to the radiators. It turns out that immediately below the skin surface of each of the radiators are countless tiny capillaries and other blood vessels. On a hot day or when an animal is overheated, temperature detectors (thermostats) in the brain sense the rise in body temperature. Signals are then sent from the brain to the muscles surrounding the tiny blood vessels in the radiator. These signals allow the vessels to relax and open up, or dilate. When the vessels dilate, more blood flows through them. The heat from the animal's blood rapidly escapes across the wall of the blood vessel, then across the skin, and finally out into the environment. Among humans, and especially in fair-skinned people, you can easily see the flushed appearance that results from the opening of all those blood vessels after a great deal of exercise. (Blushing from embarrassment is another way in which skin capillaries dilate, although it has nothing to do with a need to radiate heat.)

What about other animals? Picture a flamingo, whose legs are almost comically long and skinny. Naturally, the legs are there for walking and wading, but those are only a small portion of the bird's total activities. Therefore, the legs don't need to be especially well-developed. Long, thin legs may not have much bulk, but they do have a lot of surface area, and there are plenty of blood vessels running the length of the legs. Therefore, this is one way in which a feather-covered, warm-blooded animal can eliminate excess heat on a hot day.

Now that we've seen how bodies shed heat, let's look now at what happens in a bird or mammal on a cold day. If temperature sensors located in the skin detect a drop in air temperature, the amount of blood flowing through the radiators will be decreased. Most warm blood bypasses these areas and is rerouted to regions deep within the animal, thus minimizing heat loss. Of course, some blood must continue to feed these structures or their cells would die, but most of it is redirected. Thus, on warm days an animal's radiators are opened (turned on), and on cold days they are closed (turned off).

Mammals also have a variety of other mechanisms to keep warm on cold days. One of these is shivering, a process of spastic muscle contractions that expends energy but doesn't do any useful work. We stay in one place because we contract opposing muscles at the same time, such as those that make our legs move forward and backward. For those animals that can shiver, it represents a good way to produce internal heat. Other animals curl up into a ball to expose less body surface to the environment. We do that, too, almost subconsciously, when we're huddled under the covers on a cold night.

One of the major ways in which mammals generate internal heat is by a process known as nonshivering thermogenesis (actually, creation of heat without shivering). It requires temperature-sensitive cells in the skin to relay information to the brain, thus setting in motion two events that work hand in hand. The first event is a neural signal that causes the release of epinephrine and its related compound, norepinephrine. These compounds constrict the blood vessels of the skin, so that warm blood doesn't get a chance to reach the surface of the body, where its heat would be radiated out to the environment. They also cause the activation of a kind of fat tissue called brown fat. In animals that have it, such as bats, rodents, and newborn babies, brown fat is not

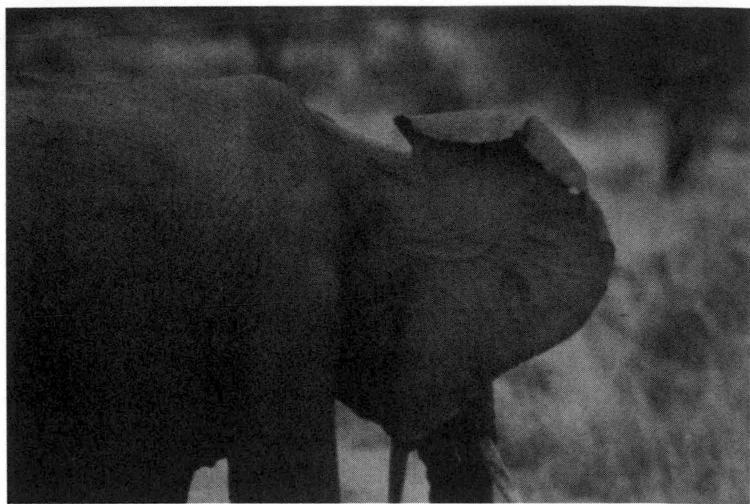

*All warm-blooded animals need to retain heat on cold days and dissipate heat on warm days. One way of doing so that is common to most birds and mammals is to regulate the amount of blood flowing to regions that are exposed to the air, like the ears of elephants (shown above) and the wings of bats (at right). Structures like these tend to have a great deal of area for heat exchange, with a highly dense system of blood vessels just below the skin surface. By flapping its ears on a hot day or beating its wings, respectively, elephants and bats can increase*

used as a storage depot for fuel, as is ordinary white fat. Brown fat (which gets its name from the high amounts of intracellular organelles called mitochondria) is actually a form of heat storage. Epinephrine activates the metabolism of the brown fat cells to begin burning glucose within the cell cytoplasm. In fact, the metabolic activity of the cells increases from nearly nothing to a very high rate in a matter of minutes. Anytime a cell begins burning fuel in this way, heat is generated, but in this case, a nifty thing happens.

At the same time that epinephrine activates brown fat cells, the hypothalamus sends a signal to the pituitary—and from there to the thyroid gland—to begin

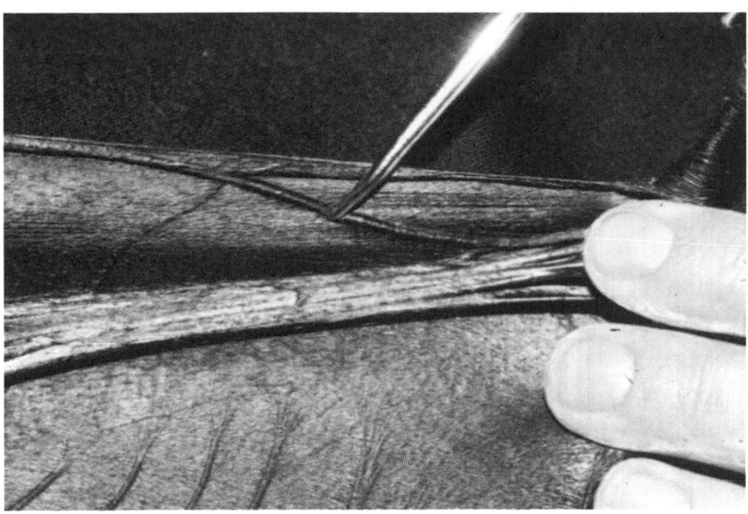

*the amount of air moving over these membranes, thus increasing the rate of heat loss. People use this same method of regulating body heat. On a hot day we feel flushed because the blood vessels in our skin open up, allowing heat from our blood to radiate outward through our skin into the air. On a cold day, we decrease the amount of blood flowing to the tip of our nose, our ears, and the skin of our arms, thus retaining heat. (Elephant photo courtesy of Dr. Charles K. Levy; bat wing photo courtesy of Dr. Thomas H. Kunz.)*

releasing higher amounts of thyroid hormone, which also acts on brown fat. One of its actions is to somehow uncouple the glucose-burning enzymes from their ability to make ATP. In short, the burning of fuel in the fat cells becomes extremely inefficient, and, like an inefficient engine, this generates more heat and less work. But that's exactly what is needed—more heat. This heat enters the circulation and helps warm the entire animal. Many newborn mammals have brown fat, because that is the time they are most susceptible to the cold.

Thyroid hormone, which is found in all vertebrate species, has many important roles in physiology. It can also increase heat production in another way, by increasing the inherent rate of activity of a key enzyme found in virtually every living cell, known as the sodium-potassium molecular pump. When this pump works overtime, it depletes cells of internal ATP, because ATP is required to drive the pump. As ATP levels decline in the cytoplasm, a signal is generated to automatically speed up the rate at which additional glucose is converted into more ATP in an effort to replenish the energy stores. This, then, generates heat as well. In fact, in some species up to 70 percent of the total heat produced by the body results from the operation of this one "pump" molecule. Not surprisingly, one of the common complaints of people suffering from hyperthyroidism is that they feel warm a lot.

## Oil and Lard: Life in the Cold

Thus, warm-blooded animals like ourselves can tolerate very low environmental temperatures, as long as we're coated with an insulating layer of fur, hair, blubber, or clothes to conserve the heat that our bodies generate internally. What about cold-blooded animals? Can they cope in a climate that is extremely cold? In general, they have a pretty tough time doing so. Fish, however, have evolved some intriguing adaptations to keep from freezing in addition to the previously mentioned antifreeze. When the water temperature around a fish drops, the fatty substances that make up about 25 percent of an animal's body weight undergo a fascinating change. The fats switch from being saturated to unsaturated.[2] To picture what happens, think of common everyday cooking oil. At room temperature it exists as an oily liquid, but if it were greatly cooled down, the oil would freeze and take on the consistency of lard. This kind of oil is called an unsaturated fat and is common in plants. Unsaturated fats stay liquid (oily) even at temperatures near the freezing point of water. Saturated fats, on the other hand, are found in animal flesh and tend to be solid at the temperature of our bodies. Saturated fats melt and become oily only if heated to very high temperatures. This is obvious when we cook animal flesh; a cheeseburger starts out as a solid, but after cooking, many of the saturated fats melt and the frying pan fills up with grease. If allowed to cool, the grease returns to its usual solid form.

Fish, like all animals, tend to have lots of saturated fats in their body cells, but if the water temperature becomes very low, a fish's body temperature will also fall. To prevent itself from turning into lard, special temperature-sensitive enzymes convert saturated fats into unsaturated ones. Once this is done, the cells within the fish can remain in their usually oily state.[3]

Finally, what about those species that can change their body temperature from warm- to cold-blooded, depending on the time of year, such as those animals that hibernate or enter into torpor? True hibernators, such as many species of bats, ground squirrels, marmots, and other mammals, are all warm-blooded. (Bears, often thought to be hibernators, are in fact not true hibernators—they lower their body temperature a few degrees, but nothing like the aforementioned species.) When winter approaches, these animals can lower their internal body temperature to just a few degrees above that of the environment. This occurs quickly and, just as important, is reversed quickly (within one day) when the outside temperature warms up. Torpor is like a mini-hibernation that may occur for only one night at a time. Many species of rodents and bats, for example, enter torpor whenever the temperature drops suddenly or when food availability is low.

The advantage of hibernation is simple. By entering a state of low body temperature, chemical reactions are slowed down and much less energy is required to support those reactions. It's a good thing, because food is usually scarce during the winter, and it may be impossible to fuel the energy demands required to support the high metabolic rate of such small animals. Indeed, many mammals that do not hibernate but live in cold climates die of starvation during the winter months. Remarkably, some hibernators can live at a very low body temperature for up to six months, waking up only once or twice the entire time.

Surprisingly, our knowledge of heat radiators in the animal world has provided a clue as to whether or not dinosaurs were cold- or warm-blooded. For years, scientists assumed that the large triangular plates on the back of a stegosaurus were there for defense purposes. This may very well have been true. It's easy to imagine such bony plates making it tough for a *Tyrannosaurus rex* to get a decent mouthful of stegosaurus flesh, but we can also look at those plates in a different way. Like the ears of an elephant, the stegosaurus' plates jutted out from its body surface, and each plate had a lot of surface area. To a physiologist, those plates look conspicuously like a classic heat radiator. An even better example was the dimetrodon, with a huge fan-shaped plate on top of the animal. Because only warm-blooded animals have such radiators, the conclusion is obvious. Of course, until we can bring a dinosaur back to life, we can't prove this to be true. Nonetheless, it was one of the first clues that dinosaurs may have been active, energetic animals resembling the warm-blooded animals of today.

So, the next time you visit a zoo, take a look at some of the features of the animals discussed in this chapter and all the other warm-blooded animals you see. Who would have thought that the nearly transparent wings of a bat could have so much in common with an elephant's ears?

CHAPTER EIGHT

# Sensing the World Around Us

*Life is a series of sensations connected to different states of consciousness.*

—REMY DE GOURMONT, FRENCH CRITIC, IN *LE CHEMIN DE VELOURS* (1902)

What would it be like to see the world from above in ultraviolet? Or to sense the person standing next to you by the electric field given off by his or her body? Or to taste pure water? Or even to find a bird's nest hidden in a bush by sensing the heat given off by the nestlings?

All of these things and many more are beyond our capabilities—and some would argue that it's a good thing they are. If we had the above abilities, we could easily be overwhelmed by our senses. But before we can explore how our sensory apparatus compares to those of the rest of the animal kingdom, we need to rethink our notion of what exactly a sense is in the first place.

From our human perspective, we are aware of five senses—sight, sound, smell, taste, and touch. As diurnal animals, we rely primarily upon sight to learn about our environment. Nocturnal animals, on the other hand, rely primarily upon smell and sound. But there are animals that can detect cues that are invisible or undetectable to us. Sharks can detect the electric field generated by the beating of a crab's heart and use it to home in on the animal even in murky waters. Some birds can see polarized light. The way we hear sound, by the bending of special "hair" cells in our middle ears, is similar to the way fish detect vibrations in water with a structure along their sides known as the lateral line.

Thus, for a physiologist, lumping all these different sensory abilities into five distinct senses becomes difficult or impossible. It's much easier to study different senses if we group them according to the type of stimulus they recognize. For instance, olfaction (smell) and gustation (taste) can both be considered chemical reception—in each case chemicals are recognized by specific odor or taste receptors. Further, all animals have within them specialized chemical receptors that detect changes in blood levels of vital substances, like sodium, glucose, and oxygen. And even single cells—like white blood cells—can detect the presence of a chemical stimulus from bacteria and orient themselves to attack it. Thus, we can consider these internal sensory structures to fall into the same category as smell and taste.

In much the same way, we can define a group of senses as those that operate by mechanical reception, or mechanoreception as it's usually called. This group includes any stimulus that results in the physical bending or deformation of a receptor cell. These may occur on the surface of the skin (at the base of hairs), deep within the skin (pressure receptors), in the middle ear (sound), in the inner ear (balance), in the lateral line of fish (vibration and touch), in the tendons and ligaments of joints (positional sense), and in all the stretch receptors found throughout the body, such as those that detect when the bladder is full or when blood pressure is high.

Yet another category of stimuli includes those that detect electromagnetic radiation and includes vision (photons), magnetic sensing, and electrical sensing. And last, thermal reception is the ability of an animal to detect different degrees of heat. Incidentally, cold is really the detection of a lack of heat; technically, there is no such thing as cold, only more or less heat.

By grouping the different types of sensory stimuli this way, it is easier to make generalizations about how seemingly unrelated senses work. For example, once it is understood that smell and taste mediate responses to the same type of stimulus—namely a chemical one—it becomes clear that the biochemical mechanisms that transduce the signal (odor or taste molecule) into perception (good or bad smell or taste) are probably either the same or very similar. This simplifies things enormously for an investigator interested in learning how senses work—and why they sometimes don't.

## Responding without Thinking

One thing that becomes clear as a result of such analyses is that most sensory stimuli are linked to a biological response—often a reflex. The best way to appreciate this is to consider why senses evolved in the first place. In the simplest organisms, there is no central nervous system comparable to our own. Conscious thought is not possible, nor is learning and memory. Thus, in order to respond to potentially significant changes in its environment, such a primitive animal would need built-in reflexes programmed to operate in response to external cues. If you shine a light on half of a petri dish containing the black flatworm planarian, for example, the worm reflexively turns around and swims away from the light. It does not think about this action, nor is it responding to the light any differently from the way our knee jerks when tapped by a doctor's hammer. It is pure reflex and designed to keep the black worm out of the light where it would be more conspicuous to predators looking for a meal.

Animals much more complex than worms also have reflexes that are triggered by sensory input. It would not do a rodent much good if it could detect the shadow of an eagle flying overhead, unless that stimulus caused the rodent to stop moving and decrease the chances of being spotted. Rodents that evolved the ability to detect the shadow but not to respond to it became meals; those that evolved the ability to play possum lived to pass on their genes. Thus, successful species were those that evolved both the ability to detect the outside world and the ability to respond in the correct way to different stimuli.

There are really only two kinds of stimuli to which a given animal must respond—those that signal a favorable circumstance and those that signal trouble. Examples of favorable cues are the odor of a potential mate, the taste of nutritious food, and the sound of a newborn's cry. Cues that signal danger are the odor of a potential enemy or predator, the taste of spoiled or poisonous food, and the alarm cry of frightened members of the same species. If an animal has little or no forebrain with which to think about each of these cues, it had better be able to respond to them automatically and immediately. Even in primates this rule still holds. If one monkey cries out in terror over something it's spotted in the distance, the other monkeys don't sit around pondering what might be wrong; they scatter immediately into the trees for protection. This is not a true reflex, the way it would be in lower animals, but it works as if it were.

We are all aware that reflexes are an integral part of our own lives. If you accidentally put your hand on a hot stove, you pull it away immediately without thinking about it. If someone were to unexpectedly light a firecracker nearby, you would nearly jump out of your skin (a remnant of the startle reflex of babies). These are extreme examples, but our lives are also governed in large part by countless subtler reflexes. The ability to stand up on a moving train is a great example of our sensory reflexes at work. As the train sways to and fro and stops and starts, we don't fall down because tiny stretch receptors in our leg muscles are activated each time an individual muscle is stretched. The instant a leg muscle is stretched due to the swaying of the train, these receptors send signals to the spinal cord, which then send return signals to other muscles in the leg, which counterbalance the stretch of the first group of muscles, helping us to right ourselves without even being conscious of it. This special class of mechanoreceptors are known as proprioceptors and are responsible for telling the brain where our limbs are in space. They are also the basis of the drunk driver test, in which an intoxicated individual is instructed to place his fingertips on the tip of his nose while keeping his eyes closed.[1]

Other subtle reflexes occur in people all the time. The nursing mother who finds herself leaking milk when she hears her infant crying in another room is an example of a reflex that begins with a sound stimulus, then is transduced into a hormonal signal that results in milk let-down. The opening and closing of the pupils when the eyes are exposed to different intensities of light is another example; photons of light are sensed and transduced into a mechanical response that contracts or relaxes the muscles of the iris and therefore helps us adapt to low or high light levels. Mechanical irritation of the nasal mucosa triggers a sneeze, which prevents inhaled irritants from penetrating down into the lungs where they could do great damage. The release of gastric acids into our stomach when we see and smell a succulent steak on the grill is an example of a type of reflex that prepares us for digestion.

## How Do Sensory Cues Trigger Responses?

From the above, we can conclude that the ability to perceive and respond to changes in the environment was one of the most vital evolutionary advances that made complex life possible. Not surprisingly, many scientists are interested in the mechanisms by which different sensory cues elicit their responses, especially since people with brain and spinal cord damage sometimes lose these abilities. The first steps by which the stimulus is received are different for different groups of senses, but the final steps are all identical.

For mechanoreceptors—like those in the ear—the process begins when a physical stimulus bends the hair-like projections on top of a sensory cell, leading to changes in the electrical properties of the cell. In the ear, the physical stimulus is a wave of fluid pressure set up by sound waves beating upon the eardrum. Different frequencies and intensities (or decibels) of sounds cause the eardrum to vibrate accordingly. This difference is amplified within the middle ear by special bones (the tiniest ones in the body) that clang against each other with each vibration. As the bones vibrate, they initiate a wave of pressure in the fluid that encases the structures of the middle ear, and this bends the hair cells.

For nonmechanoreceptors, the usual way in which a stimulus is perceived is by the binding of a stimulus molecule to a protein receptor molecule on the sensory cell. In the nose, for example, a molecule of, say, perfume dissolves in the mucus of the nose and then binds in a lock-and-key fashion to a specific receptor on sensory cells that project from the brain into the nasal cavity. This binding reaction changes the electrical properties of the receptor cell, just as bending the hair cells did in the ear, eliciting in the brain the pleasant smell of the perfume.

One question that arises from this is that of the specificity of the stimulus. Why is it, for example, we can smell odors that never existed at the time mankind was first evolving the sense of smell? Take, for example, modern day chemicals synthesized in a laboratory, including such things as synthetic perfumes. We surely didn't evolve a receptor protein to detect synthetic organic compounds like dichloromethane—a commonly used solvent—but I assure you we can smell it. Likewise, we can taste artificial sweeteners and other synthetic foods and drinks. These and other examples indicate that the receptors for such odors and taste molecules must not be terribly specific, that is, a "peppermint" receptor might preferentially detect molecules of peppermint but might also detect a host of other odors as well (albeit to lesser extents). This would also be true of all the several thousand receptors in the nose. Thus, the shape of a molecule of Chanel No. 5 might have just enough structural similarity to dozens of other natural compounds that it is able to fool each of the relevant receptors into binding it, at least to a limited extent. What happens next is up to the brain. The brain must integrate all of this new information and perceive the stimulus. In the case of Chanel No. 5, the perception is a pleasant one; in the case of dichloromethane, it is unpleasant.

Likewise, we can also see more colors than the corresponding number of receptors in our eyes. Our eyes are only built to receive the light signals of three specific wavelengths, those corresponding to red, green, or blue light. But there is enough splay in the specificity of these receptors that some light rays can stimulate two or more of the receptors. Depending on the overall ratio of receptor types that are stimulated by a light source, we perceive the signal as either orange, violet, or turquoise, or any other shade of color. For instance, a light ray that activates the red receptors of our eyes to 99 percent of their maximal excitation capacity and the green receptors to 42 percent without activating the blue receptors is perceived as orange. A beam of pure yellow light would activate equal numbers of red and green receptors but no blue receptors.

Another important feature of our senses is that the stimulus signal can often be amplified. When an odor molecule binds to a receptor, it activates enzymes in the sensory cell that then make hundreds of intracellular mediators responsible for producing the electrical changes that are required for the information to be sent to the brain. Thus, a single odor molecule can have its action amplified hundreds of times. Sometimes amplification is critical in determining the source of a signal. Bats that sense their environment by the process of echolocation are a good example. Think of the odd sensation you may have experienced while driving and hearing an ambulance siren in the distance. For a few moments, it is difficult to determine if the siren is coming from somewhere behind or ahead of you, because the sound waves are reaching both of our ears at almost the exact same instant. Our brain usually enters default mode under such circumstances and assumes that the source of the sound is in front of us. By contrast, if someone were to speak to us from across a room, we'd have no trouble identifying where they were even with our eyes closed. Under those circumstances, the slight difference in arrival time of the sound waves on each side of the head would tell the brain that the signal reached one ear sooner than the other—hence, the signal must be coming from that side. Most animals that can hear at all have this ability.

What does this have to do with echolocating bats? In order to capture an insect, a flying nocturnal bat emits high frequency sound—well beyond the audible range of humans—and detects the returning echo of those sound waves as they reverberate off the insect. This tells the bat where the insect is and also how big it is. But a bat has a tiny head, and thus its ears are very close together; sound waves will therefore reach both ears almost simultaneously. This would make it very difficult for a bat to locate the source of its food were it not for a built-in amplifier in its head. As the signal reaches one ear, inhibitory electrical signals are sent to the neural circuits originating from the opposite ear. These inhibitory signals slow down the rate of transmission of signals from the far ear to the auditory center of the brain. Thus, the slight delay in arrival time to each ear is amplified into what appears to the brain as a longer delay—one that can more readily be detected and acted upon.

Directionality is important for most animals, and that's why animals tended to evolve two sets of receptors for each stimulus. Fish have lateral lines on both sides of their bodies so that they can compare vibrational signals and determine the source of the vibration. A wrong decision could mean swimming right into the jaws of death. Certain reptiles have extremely sensitive thermal receptors called heat pits on either side of their head; these allow them to determine the source of warmth emanating from a nearby warm-blooded animal (for example, a mouse). We have two ears for the reasons stated above but also two eyes for depth perception and two nostrils. Have you ever wondered why we have two nostrils instead of a single hole in our nose? It would seem to be easier to inhale air through one large opening than through two smaller ones. Perhaps we and other animals evolved two nostrils so that we could better detect the source of an odor. Just like a snake, whose forked tongue samples the air for odors and uses the slight difference in odor concentration on each fork to determine the direction of the smell, at one time we may have relied more heavily upon our olfactory abilities.

Nowadays, smell is our least important sense. (In general, women have a keener sense of smell than men; oddly enough, it's especially acute during ovulation.) Many people go through all or part of their lives completely anosmic (unable to smell), often due to head trauma. Although this certainly reduces the quality of their lives, it has few consequences, such as the inability detect the odor of spoiled food or a gas leak in the house. To appreciate how little we rely upon our sense of smell for understanding the world around us, think of the millions of chronic allergy sufferers who are almost always congested. They certainly get by in life without any significant consequences related to their loss of smell (apart from a decreased ability to taste food, which is linked with the sense of smell).

Like humans, most birds apparently have a relatively poorly developed sense of smell. On the other hand, because birds are primarily diurnal, they have extraordinary eyesight. Thus, as their visual abilities evolved, their olfactory ones declined. Interestingly, however, some birds (like the common starling) increase their sensitivity to the odors of certain plants whose growth is synchronous with the onset of the nesting instinct. Birds that are able to smell these plants have a selective advantage. The vegetation that attracts starlings contains a bactericidal and antiparasitic agent. When this vegetation is used to line the nest, it provides a natural protection for the chicks against invasion by pathogens. Without such a lining, many of the offspring would succumb to disease soon after birth. However, once the mating and nesting period is over, the starlings lose this ability to smell again. Presumably, the change in olfactory sensitivity is triggered by the rise in reproductive hormones that precedes mating in the spring.

The process of detecting a sensory stimulus is often quite sensitive as well. For example, insects can detect odor molecules down to a few parts per million (a male moth can smell a female moth a hundred yards away), and elephants can communicate via low frequency sounds over a distance of many miles. Most animals, including ourselves, are exquisitely sensitive to even minute amounts of sugar. Without this sensitivity, our senses would be much less useful. For instance, if a gazelle needed to be within a few feet of lion before it could detect its smell, there would be a lot fewer hungry lions around.

As sensitive as the receptor cells are, however, there are times when less sensitivity is needed. Have you ever noticed that when you are really hungry you can smell the odor of food cooking from far away? Let's say you smell the hamburgers cooking at a local fast food restaurant. After you've finished eating, however, you're no longer as conscious of the smell. Your odor receptors have adapted to the stimulus and, because it's no longer important (at least for the time being), they ignore it. This helps us focus on other sensory cues that now assume more importance than olfaction. Likewise, you may be acutely aware of a bad smell when you first encounter it, but after awhile you don't notice it as much. Similarly, the drone of an air conditioner becomes a sort of white noise after awhile, and we are aware of a hat on our head when we first put it on but soon after don't feel it as much. The same is true of your backside when you sit on a chair for a long period. These sorts of adaptations occur at the molecular level in sensory cells and prevent us from being bombarded by sensory stimuli. Can you imagine how awful life would be if you were consciously—and constantly—aware of each sound, sight, smell, taste, and touch signal you were receiving! Adaptation of sensory cells to chronic input is, therefore, an important survival mechanism that evolved early in almost all animals.

## How Animals See the World

Although each sensory stimulus probably activates receptor cells in similar ways in most vertebrate animals, that does not mean that all animals perceive the same things. Unfortunately, we can never truly know what an animal sees, hears, or feels, because most of

*The world looks very different through the eyes of other animals. For many years, scientists wondered why certain nondescript-looking flowers attracted insects, even when there were more vividly colored flowers in the same vicinity. When it was discovered that some species of insects (and even vertebrates, like hummingbirds) have special photoreceptors in their eyes that are sensitive to ultraviolet light, the mystery was solved. In these illustrations, the same flowers were*

what is perceived is the result of the workings of the brain. But we can make educated guesses, based on our own experience and on experimental recordings from animals. For example, we know that birds have at least one, and possibly two, additional light receptors in their eyes, giving them an extra degree of complexity in establishing ratios of receptor stimulation. Their brain undoubtedly receives more complex ratios than our brain. Do they see colors we can't even name? Does the world look entirely different to them? Certainly, some species of birds (for example, hummingbirds) can see light in the ultraviolet region that is too high in energy

*photographed in normal and ultraviolet light, the second photo revealing a striking visual pattern normally invisible to us but not to many other animals. We can only try to imagine how the world looks, smells, tastes, and feels to other animals. For that matter, no one person can really be sure if the way we perceive the world is exactly the same as anyone else. When you see the color red, do you see the exact same degree of redness as someone else? (Photos courtesy of Dr. Thomas Eisner.)*

for our eyes to detect. Many insects can do that, too. We know all this because we can isolate the receptor cells, called photoreceptors, and test their responsiveness to light of specific wavelengths. Using special photographic techniques, it's become clear that many objects in nature—such as certain flowers that appear colorless to us—are actually beautifully patterned in ultraviolet images.

There exists a whole visual world out there that we are not privy to. What must it be like to view the world from the perspective of a bird of prey, like a falcon? During a dive-bomb attack on another bird, a peregrine falcon can reach speeds of 150 miles per hour but still focus on its prey. When we travel at even half that speed in a car or train, we lose the ability to focus on objects unless they are far in the distance. Likewise, a falcon can see the individual pulses of light coming from a bulb operated at 60 Hz (cycles per second); to us, this looks like a constant beam, but to the bird it looks like a strobe light. Likewise, the spokes on a moving wheel are spinning too fast for us to separate them, but a bird sees the individual spokes even when they appear as a blur to us. Thus, the temporal resolution of birds is much more extraordinary than ours. This is a necessity for an animal moving quickly and hunting on the wing. Incidentally, it is believed that the reverse is also true, that is, birds can detect slowly moving objects better than us, such as the motion of the moon through the sky. Birds of prey also have larger, more complex eyes than ours, which allow them to capture more light and see at a greater distance. Their eyes are similar in some ways to a telescopic lens on a camera; images on the ground, such as a moving lizard or rodent in the brush, are magnified as many as three times by the bird's eye. The structure of their retinas also suggests that what they see is not only magnified but is a sharper image as well.

A bird's eyes are so important to their sensory perception that they tend to be a great deal larger than for most other animals. In fact, the largest eye in the animal kingdom doesn't belong to whales or elephants but to ostriches. Eyes may occupy up to 50 percent of a bird's cranial cavity (compared to 5 percent for us). To be comparably sized for our heads, our eyes would need to be about the size of baseballs. Unfortunately, this doesn't leave room for a lot else (hence the term "bird brain"), and the eyes of some birds take up so much room that there is little left for muscles to attach to the eye. These muscles give us the ability to move our eyes around independent of our heads; most birds have to move their entire head if they want to see what's around them.

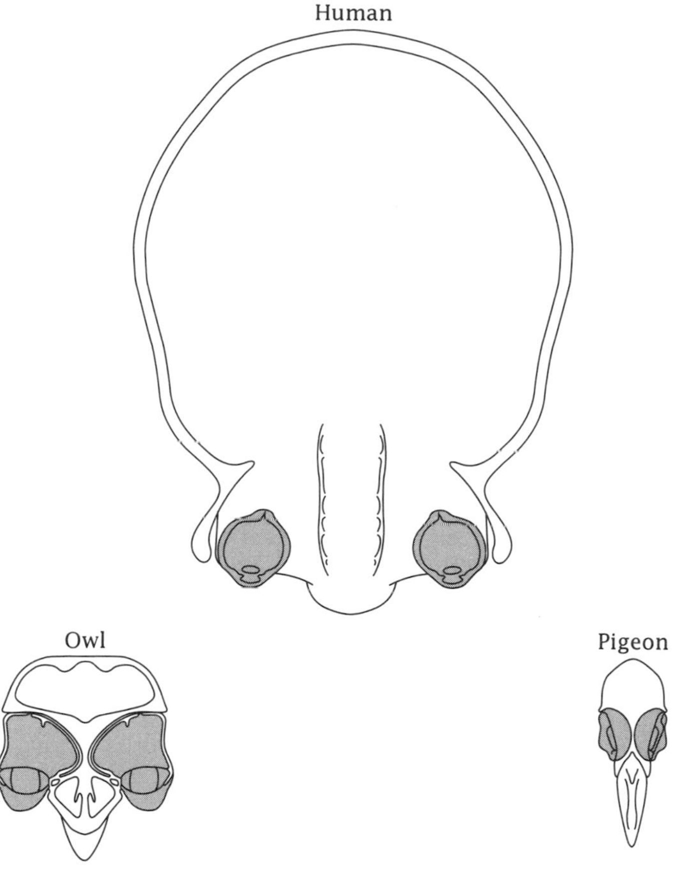


*Birds, such as owls and pigeons, have extraordinarily good vision, in part because the eyes have so many more light-capturing cells, or photoreceptors. To make room for all these photoreceptors, the inner structure of the eye of a bird has become quite enormous, occupying about 50 percent of the skull cavity, as compared to the eyes of a person, which take up only about 5 percent of the skull. The largest eye of any animal belongs not to a whale or an elephant, but to the ostrich. Note also that the eyes of man and owls face forward, for good depth perception. The owl is a hunter, and our eyes face forward because we evolved as arboreal animals (depth perception is obviously vital for animals that swing from branch to branch). Pigeons and other nonhunting birds have eyes placed on the sides of their heads. What they sacrifice in depth perception, they make up for in field of vision, making it easier to locate predators attacking from any direction.*

Having big eyes, however, means better light capturing ability and more retina for light to focus on. The retinas of birds not only have additional types of color receptors but also more receptors packed into a given area than is found in mammalian eyes. This creates a sharper picture, just as adding more pixels on a computer screen gives a sharper image.

As a rule of thumb, birds of prey (in fact, predators in general) have forward-facing eyes like ours; this gives added depth perception and is especially important to hunters that need to quickly lunge toward potential prey. Plant-eating animals—like deer—and all other birds have laterally facing eyes. In these animals, depth perception is sacrificed in return for a wider field of vision. Although their vision is monocular, having eyes on the sides of the head means it is possible to see images (like an attacking predator) around a field of almost 360 degrees. On the other hand, a deer's monocular vision probably explains in part the tendency of deer to crash through plate glass windows!

It is clear that animals also detect other senses differently than us. We know that pigs, dogs, and other animals can taste water, for example. They have special taste buds that respond to water the way our salt taste buds respond to salt or our sweet buds respond to sugar. (If you are thinking that you can taste water, you're wrong. What you're actually tasting are impurities and chemical additives in the water.) For all we know, water may taste absolutely delightful to them, and in some ways it is surprising that the ability to taste water isn't universal. If there is anything you would assume would taste delicious to all animals, it would be water because of its role in keeping us alive.

Some animals can taste table sugar (sucrose), some can taste artificial sweeteners like saccharine, some can taste both, and some taste neither. Our sensitivity to sweetness is well known and is something most of us sometimes wish we could do without. It's an example of a survival adaptation that has outlived its usefulness. We probably evolved such exquisite sensitivity to sugar to encourage us to eat foods like fruit that contain low amounts of sugar. Sugar is a necessary nutrient and provides energy but in excess results in weight gain. Our sensitivity to low amounts of sugar and our ability to cultivate sugarcane have resulted in taste buds that have grown accustomed to being flooded with sugary foods; we've become jaded to the sugar in natural foods.

Although it may seem odd that we have no ability to taste water, one might equally wonder why the ability to taste food exists at all. While one obvious reason would be to encourage animals to eat the things their bodies need, this was probably a later development. Actually, the sense of taste may have evolved as a means to warn animals off poisonous foods, because many poisons in nature have a bitter taste. Most animals will immediately spit out something with a bitter taste, thus unwittingly protecting themselves from serious danger. Later, taste buds specific for salt and sugar evolved to ensure that these vital compounds were eaten. The sour taste buds, which actually detect acidic compounds, may have evolved to give us the ability to taste citrus foods, which contain important nutrients like vitamin C. If animals and early *Homo sapiens* did not have taste buds for such compounds, it's highly unlikely that they would have begun eating the diverse diet we all need to survive. Thus, taste buds are not simply something that provides us with a pleasurable sensory experience, but are actually the result of important evolutionary pressures that contributed to survival.[2]

## How Animals "Feel" the World

While we can relate to the senses of sight, smell, sound, touch, and taste, there are other senses or perceptions in the animal kingdom to which we cannot possibly

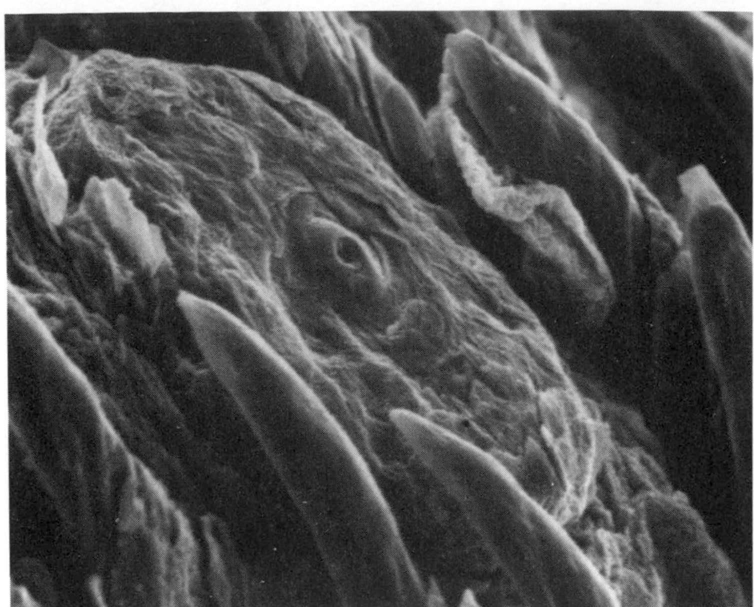

*This bizarre image is a scanning electron micrograph of the surface of the tongue. The round structure in the middle is a taste bud, surrounded by hair-like projections that contain the chemical receptor molecules for identifying different tastes. Human beings can recognize at least four discrete tastes: salt, sweet, bitter, and sour/acid. Some animals can recognize other types of taste, such as that of pure water, which we cannot detect. (Given its survival value, it's surprising that we cannot. How it must taste to a dog is anyone's guess, but presumably it tastes very good.) The ancient Chinese discovered that an extract of seaweed could enhance the flavor of foods, even though it had no taste of its own. This extract was identified in the early twentieth century as monosodium glutamate and is now processed from wheat gluten and marketed as a flavoring agent. Glutamate is a potent stimulant of nerve cells, including those that surround taste buds. Thus, adding MSG to food increases the likelihood that a taste bud will be stimulated, even at low "flavor" concentrations. The use of MSG in Chinese restaurants has led to the identification of what is known as Chinese Restaurant Syndrome; glutamate can cause painful vasodilation of the blood vessels in the brain, creating nasty headaches and other symptoms.* *(Photo copyright ©SIU/Peter Arnold, Inc.)*

relate. What must it feel like to a shark as it swims through dark, murky waters and monitors the electrical signals generated by other animals? Imagine feeling the electrical aura of each person or animal you encountered every day. What would that feel like—a tiny shock, perhaps? A faint vibration of some sort? A tingling sensation? It's hard to assign human terms to something with which we have no experience. Even harder is trying to imagine what the Earth's magnetic field feels like to a pigeon or hummingbird. There are tiny magnetic particles in the head of a pigeon that change their orientation as the bird travels through the earth's magnetic field. But how would that feel? Like someone knocking us on the head? A slight feeling of pressure? Dizziness? Is it pleasurable or painful? One of the eternal frustrations of comparative physiologists is that we can determine *what* an animal feels but not *how* it feels.

The first steps in interpreting a stimulus are similar in many ways for all animals. They must detect the stimulus with a sensitive receptor, adapt to it if the stimulus is constant, amplify the signal if needed, and determine its direction. To perceive and exactly interpret what the stimulus is, however, electrical signals must be sent to the brain. This is where things become a little cloudy.

Every stimulus, whether a sound wave, light wave, odor molecule, vibration, or taste, results in the generation of an electrical signal called an action potential. Action potentials arise when a sensory cell is activated. Often the sensory cell is itself a neuron, like those that detect odors in the nose. Other times—as in taste perception—the sensory cell is not a neuron itself but is closely associated with one and activates it when a signal is received. Astoundingly, however, the electrical signal coming from a pain fiber, a taste bud, or from the nose are all exactly the same—simply electric current. Thus, there is nothing inherently "smelly" about smells or "painful" about pain. The reason we see a sight or smell a smell is that the action potentials terminate in specific regions of the brain devoted to interpreting those signals. If a pain signal were somehow rerouted to the visual cortex, we would "see" pain, not feel it. Likewise, we might "hear" a sunset if the nerves from our eyes were linked to the auditory cortex instead of the visual centers. It's an odd way to think about the senses; but it is absolutely clear that the work of determining the qualities of a given stimulus occurs within the brain.

This remarkable ability of the brain is what allows us to smell new odors, like perfume, to see additional colors than the three our eyes are built for, and to determine how intense a stimulus is. It also allows us to distinguish between the feel of a feather on our back and a needle pricking our skin. However, this central processing design means that if the brain centers that interpret a given stimulus are destroyed, the ability to detect that sense is also lost. This is why brain injuries can result in partial loss of sensory ability, even though the sensory cells are themselves unaffected. Some of these losses lead to rather bizarre sensory experiences, like hallucinations.

Clearly, life without sensory stimulation would be dull indeed. Our brains are constantly sorting out relevant from irrelevant cues and charting an appropriate response to those cues. Some stimuli are pleasant, others are aversive, but it's the ability of such stimuli to elicit reflexes and other behaviors that permits an animal to survive in a world full of lurking dangers and unexpected pleasures. In the next chapter we'll explore what happens when the status quo is interrupted—when the usual challenges to survival become too intense. How does the sensory system interact with the other systems of the body to restore homeostasis in life-threatening situations? And do we respond to such challenges any differently than our prehistoric predecessors?

## CHAPTER NINE

# Stone Age Stress and Coping with Change

*War will never cease until babies begin to come into the world with ... smaller adrenal glands.*

—H. L. MENCKEN, JOURNALIST, *MINORITY REPORT: H. L. MENCKEN'S NOTEBOOKS* (1956)

What happens when someone falls through the ice in winter, loses blood after a car accident, or develops pneumonia? In these situations, where the external environment (the temperature of the icy water) or the internal environment (the amount of blood in the circulation or a bacterial infection) are changing, we are no longer considered to be in homeostasis. Some years ago, the brilliant Canadian anatomist Hans Selye used the word "stress" to refer to such situations and any others that result in what we now call the internal stress response. Thus, the word "stress" has a true working scientific definition, in addition to the more colloquial usage with which most of us are all too familiar.[1]

Nowadays, however, we are bombarded with what I call the mythology of stress, which suggests that our psychological and physiological well-being is constantly threatened by degrees of stress unparalleled in history. Nothing could be further from the truth. It may be true that many other species of animals are more stressed than their primitive ancestors, because of environmental pollution, reduction of territories by large-scale clearing of forests, and other human-inflicted catastrophes. On the other hand, the modern person living in a developed country like the United States is, if anything, living a more mellow life than his or her primitive ancestors.

What are some of the real or imagined types of stress that people battle today? Coping with rush-hour traffic, job and financial difficulties, troubled relationships, family problems, school responsibilities, and mild illness are just a few of the many stressful stimuli that can be identified. Anxiety over personal problems (will I be able to pay the rent this month?) or more global concerns (will there be another war?) is another type of stress that we encounter all too often. Nonetheless, anxiety and these other stressors are not immediate threats to survival, even if they do occasionally make our blood boil. Of greater concern is that the internal defense mechanisms of the body respond to these less critical types of psychological stimuli in the same way they respond to actual life-threatening ones. Why is this unfortunate? Because over the long haul, an excess of potent stress-fighting factors, like the adrenal gland hormones cortisol and epinephrine, can suppress the immune system, contribute to ulcers, produce muscle atrophy, elevate blood sugar, place excessive demands on the heart, and may even lead to death of certain brain cells. Thus, an individual in the midst of a divorce does not require the same hormonal, neuronal, and metabolic responses as someone who falls through thin ice on a pond in winter. In both cases, however, the same internal changes are initiated.

Why then do emotionally stressful events elicit the same chemical changes in our bodies as events that are actual threats to survival? The answer may lie in a comparison of stress as we know it today and stress as it must have been when vertebrate animals were first evolving. Are we really any more stressed out than our prehistoric ancestors? That's hard to say, of course, because we can only speculate on what life was like for primitive man. Stress, however, is clearly not a recent phenomenon. For instance, the defense mechanisms that developed in mammals, including ourselves, did so very early in the evolution of life. We even see similar biological responses to stress in nonmammalian vertebrates, like birds, reptiles, and fish. These defenses consist of hormonal and neuronal signals that increase breathing, accelerate heart rate, increase blood pressure, increase the liver's ability to pump sugar into the bloodstream, and open up blood vessels in the large muscles to maximize nutrient and oxygen delivery. The increased cardiovascular and respiratory activity during stress has traditionally been called the fight or flight response, the idea being that such adaptations prepare an animal for the eventuality of fighting (and all it implies, like injury and loss of blood) or fleeing (and the muscular work it requires).

The term "fight or flight" was coined by the American physiologist Walter Cannon (1871–1945), who actually concentrated most of his work on understanding the gastrointestinal tract. Nearly forgotten today is the fact that Cannon was the first to use radio-opaque compounds like bismuth, together with X rays, to visualize abnormalities of the intestinal tract, a procedure still used today. But he's remembered now for his concept of homeostasis and the role of the adrenal glands in that respect. Like Galen nearly 2,000 years earlier, Cannon received part of his training in the military. His observations of wounded soldiers on the battlefields of World

War I were grim demonstrations of circulatory shock and led in part to his ideas about stress responses.

The fight or flight response is mediated primarily by epinephrine, formerly known as adrenaline (the word "adrenaline" is derived from the Latin words *ad* and *renis*, meaning "toward the kidney," because the adrenal glands sit atop the kidneys; today we say epinephrine, from the Greek *epi* and *nephros*, or upon the kidney).

If you've ever had a burst of epinephrine in your system, you will know well the classic signs—trembling hands, anxiety, jumpiness, cotton mouth, racing heart or "palpitations," and dilated pupils. The result is an animal that has lots of fuel in its blood, a more forceful heart to pump the blood around, plenty of oxygen, and efficient muscles. For an antelope in the wild that has spotted a nearby lion, these changes are exactly what the antelope needs to avoid becoming a meal. Similarly, a woman who sees a child trapped under a car needs her body to respond to this type of stress, to enable her to summon the superhuman strength needed to lift the back of the car and rescue the child.

Not surprisingly, animals evolved built-in mechanisms to combat the stresses of infection, starvation, dehydration, pain, and lack of available oxygen to breathe, to name just a few. To get the ball rolling, however, the brain must be involved in the stress response. The primitive regions of the brain known as the limbic system and the brain stem are directly linked with another ancient region, called the hypothalamus, which, you may recall from Chapter 3, was shown to be the seat of our hunger-regulating mechanism. Emotional stresses are relayed to the hypothalamus via the limbic system. Physical stresses are relayed via the brain stem. Once the signals converge on the hypothalamus, neurons are activated to produce a substance called corticotropin-releasing hormone, or CRH, a molecule that initiates the fight or flight response. In addition, CRH activates a complementary hormonal response that further prepares an animal for stress. It does this by stimulating the release from the pituitary gland of adrenocorticotropic hormone (ACTH). ACTH has only one job—to activate the outer zones of the adrenal gland (known as the cortex) to produce cortisol (or, in some animals, its chemical cousin corticosterone). Cortisol and corticosterone are known collectively as glucocorticoids, and their actions are widespread and profound. They are members of the steroid class of hormones, because they are derived from cholesterol.[2] Cortisol is one of the most potent known endogenous chemicals capable of breaking down components of the body into smaller, usable chunks for energy production. This type of action, whereby a compound destroys certain tissues in the body to provide fuel, is known as catabolism and in measured doses is vital to survival. Bone, muscle, lymph, fat, and other tissues are catabolized to provide chemicals that can serve as substrates that the liver can convert into sugar (glucose). Thus, the glucose is essentially formed by the body's own self-digestion and can supply the extraordinary demands of the heart and brain during a crisis. Fatty acids, liberated from fat stores by cortisol, also act as energy carriers for most parts of the body during stress. In any stressful situation, it is easy to imagine that extra energy would be needed, and thus glucose and fatty acids rise quickly in the blood in response to stress.

Stressful inputs can originate within or outside of the brain. Visual cues entering the eyes (a cat spotting a dog) reach the limbic system and activate the hypothalamus. Painful inputs (stepping on a nail) go from sensory nerves in the foot, through the spinal cord, into the brain stem, and finally to the hypothalamus. It's not surprising, therefore, that physical and psychological stress show the same manifestations, which is exactly what Selye first discovered, although he didn't know the anatomical basis for it at the time.

There is another interesting aspect of the stress response in mammals in addition to the generation of new energy sources. A natural painkiller called endorphin (from endogenous morphine) was developed during the course of evolution to combat severe pain. In the aforementioned antelope-lion scenario, one might envision the antelope being swiped by the lion's claw but escaping to live another day. Its endorphin would allow the animal to cope with the pain of its wound, if only temporarily, and continue with the herd. Incidentally, the discovery of endorphin in people generated a great deal of excitement in the scientific community as a potential treatment for people with intractable pain. The idea was that endorphin could take the place of pharmaceutical (and often addictive) painkillers. Unfortunately, endorphin turns out to be itself very addictive, and its potential use as a painkiller is questionable.

Other hormones, such as antidiuretic hormone (or vasopressin),[3] enable the kidney to retain more water than normal during periods of drought and dehydration or during hemorrhage, two other kinds of life-threatening stress. All of these varied measures are short-term responses to different types of stress, but they act in a concerted way to give an animal a fighting chance to get back on its feet.

What about the type of stress that is not life-threatening but is perceived to be of potential danger? When the antelope spotted the lion, there was not yet any physical damage to the antelope's body. Nonetheless, the hormonal systems responded as if the damage was already done, in anticipation of impending doom. If the crisis were luckily averted, a complex system of hormonal feedback loops would apply a brake on the stress response to prevent unabated secretion of cortisol and other stress hormones. This is quite important. Although the stress hormones are needed for the short-term adaptation to stress, prolonged secretion of cortisol, for example, can be very deleterious indeed. Imagine what would happen if the catabolic actions of cortisol failed to stop and our bodies went on extracting

*This zebra has survived a vicious attack by a lion because its stress response was activated in time. The endorphins released by its pituitary gland dulled the pain of this awful wound long enough to allow the zebra to flee to safety. The adrenal hormones cortisol and epinephrine increased the strength of the heart and the capacity of the lungs to power the leg muscles while running. Although it has survived, it still runs the risk of infection due to the open wound on its flank. If it should survive and reproduce, however, it will help maintain a genetically fit population of animals suited to the harsh life of the Serengeti plains. (Photo courtesy of Dr. Charles K. Levy.)*

energy from all available tissue. Perhaps the most dramatic example of the consequences of excess cortisol is found in a small Australian marsupial known as *Antechinus*. During the mating season, the level of cortisol rises in the blood of the male, which persists for a few weeks after mating. Eventually, the cortisol causes such profound immunosuppression that all the males die from disease, and that means more food and resources for the pregnant females and, eventually, their offspring. In other words, once the males have performed their function, they are expendable and are killed off by their own hormones. This seems to be a logical approach for conserving scarce resources, but thankfully it's one that never evolved in primates!

Fortunately, the process of negative feedback works very well in healthy people and prevents cortisol from rising too high. Feedback is the cornerstone of homeostasis and can be quite complex. Feedback can occur at the cellular level or the whole-body level. For example, when the energy-producing units of a cell—the mitochondria—begin converting glucose into ATP, the ATP feeds back to inhibit the enzymes that act on glucose. In this way, a cell makes only enough energy to serve its needs and saves the rest of the glucose for later. This type of feedback is called negative feedback, because the end-product of the process (ATP), inhibits the process from continuing any further.

Feedback, however, can be either negative or positive and can even switch from one to the other. An intriguing example of this happens each month in women (and every few days in smaller female mammals). Estrogen, produced by the follicle cells of the ovary, normally applies a brake on the release of the pituitary hormones that cause the follicle to grow and make a mature egg cell. These same pituitary hormones are also responsible for stimulating estrogen production from the follicle. But when estrogen levels get high enough, as they do once each month in premenopausal women, the negative feedback of estrogen on the pituitary somehow switches to a positive feedback action. Estrogen then causes not less but more pituitary hormones to be released, and these stimulate more estrogen and follicular growth, which stimulates yet more pituitary hormones, and so on. This type of positive feedback is relatively rare in nature, because there's no obvious way to stop the process once it's begun. In the estrogen example, the process stops only when the follicle finally grows so big that it literally blows up, a process known as ovulation. An egg is released from the ruptured follicle, and the follicle decays, allowing estrogen levels to decline and the pituitary to be released from positive feedback.

Perhaps the most widely applied use of the idea of negative feedback in the human population is the use of oral contraceptives. These are the only pharmaceutical drugs ever invented whose actions are so widely known that they need no other name than "the pill." The low levels of estrogen in the pill inhibit the pituitary by negative feedback, and thus the hormones that would otherwise stimulate growth of the follicle and egg are not released in normal amounts, and ovulation never takes place.

In the stress response pathway, feedback is primarily of the negative type and occurs at the hypothalamus (to inhibit CRH production) and at another brain site, the hippocampus. In fact, feedback is so important that it probably also occurs at numerous other sites as well, like the brain stem and pituitary. As a general rule, when it is critical to precisely regulate a physiological system, nature has evolved redundant mechanisms to make absolutely certain that things work the way they are supposed to.

## Stress and Adrenal Steroids: In Sickness and Health

Some investigators believe that chronic stress can be seriously detrimental to survival because the feedback process starts to decay with age. The cells of the hippocampus, which mediate part of the feedback action of cortisol, are sensitive to prolonged cortisol stimulation. After a while, the hippocampal cells begin to deteriorate, and thus the feedback inhibition of CRH becomes less efficient. Therefore, cortisol levels will remain high in the circulation after each stress, no matter how mild, for longer than if feedback was operating normally. This will expose more hippocampal cells to cortisol, which

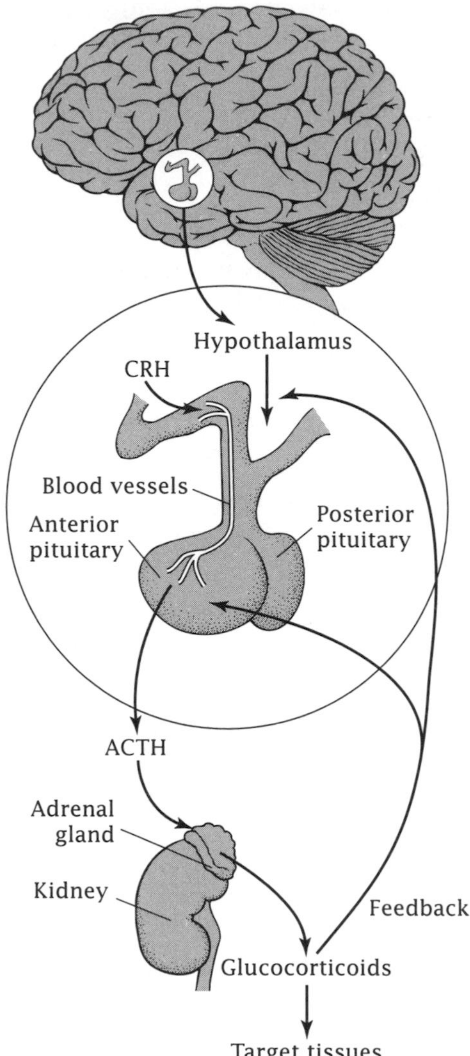


will create more hippocampal destruction, and so on. In a sense, this resembles a type of positive feedback, albeit a pathological one.

One way in which we've learned about the potential dangers of excess glucocorticoids in human beings comes from observations of people who have been

*As shown at left, the adrenal glands do not function on their own. To produce cortisol, the adrenals must first be stimulated by the pituitary hormone ACTH, which in turn is released after stimulation by the hypothalamic hormone CRH. Cortisol is extremely potent, however, and although its actions on cardiovascular, respiratory, and immune function are vital to survival, excess cortisol results in the symptoms of Cushing's disease. In that disease, bone, muscle, and immune tissue are broken down, blood pressure soars, and body fat increases. To prevent this, a complex feedback loop operates between the brain, the pituitary gland, and the adrenals. Cortisol from the adrenal glands shuts off secretion of CRH from the hypothalamus and ACTH from the pituitary. If cortisol is needed again, as during the fight or flight response to stressful situations, the feedback is overridden by additional CRH release. (Modified from Robert Sapolsky,* Why Zebras Don't Get Ulcers: A Guide to Stress, Stress-related Diseases, and Coping, *copyright ©1994 by W. H. Freeman and Company. Used with permission.)*

chronically treated with glucocorticoids for one reason or another (such as arthritis). Such individuals may show signs of immunosuppression, bone loss, muscle atrophy and other harmful side effects if the dosage of glucocorticoid is too high or therapy is too prolonged. Also, patients with tumors of the pituitary gland may produce too much ACTH, which overstimulates the adrenal glands to make excess cortisol. This disease, known as Cushing's disease,[4] results in hypertension, immunosuppression, hyperglycemia (excess sugar in the blood), and the same problems that a person develops if injected with excess cortisol.

Scientists postulate that the immunosuppressive effects of cortisol developed as a "check" on the immune system. In other words, people with the opposite problem of Cushing's disease, namely too little glucocorticoid production, tend to have a higher incidence of so-called autoimmune diseases. In such a scenario, the body's immune system runs amok and attacks the body's own cells as if they were foreign. Oddly enough, one of the primary causes of adrenal insufficiency (known as Addison's disease after its discoverer, British physician Thomas Addison) is autoimmune destruction of the adrenal glands. John F. Kennedy was perhaps the most famous sufferer of Addison's disease (although it was never acknowledged during his lifetime).

Thomas Addison (1793–1860) led a rather tragic life. He noted a syndrome of lethargy, muscle weakness, low blood pressure, loss of appetite, light-headedness, irritability, and gastrointestinal problems in a population of his patients. All of them died within a couple of years of developing this odd disorder. During autopsies of these patients, he observed that the adrenal glands were diseased in each case. He reasoned that the adrenals must produce some factor that is necessary for good health. This was well before anyone had any idea that hormones even existed. The adrenal glands were considered by many to be vestigial bits of tissue with little or no significance. In fact, this was the first time that anyone made a clear connection between a disease and any endocrine gland. We now know that additional symptoms of Addison's disease include low blood sugar and sodium and high blood potassium, each of which can be life-threatening. Addison's idea was ridiculed by the London intellectual elite, and he was forced to publish his findings on his own in 1855 (*On the Constitutional and Local Effects of Disease of the Supra-renal Capsules*). It was not until some 5 to 10 years later, however, that his ideas gained acceptance and, eventually, fame for Addison. By then, however, Addison was despondent over his treatment by the scientific societies of the day—and apparently a morose person by nature to begin with—and committed suicide in 1860, just missing what would have been his rightfully earned place in the scientific world. Incidentally, Addison was also the first person to describe the disease known as pernicious anemia,

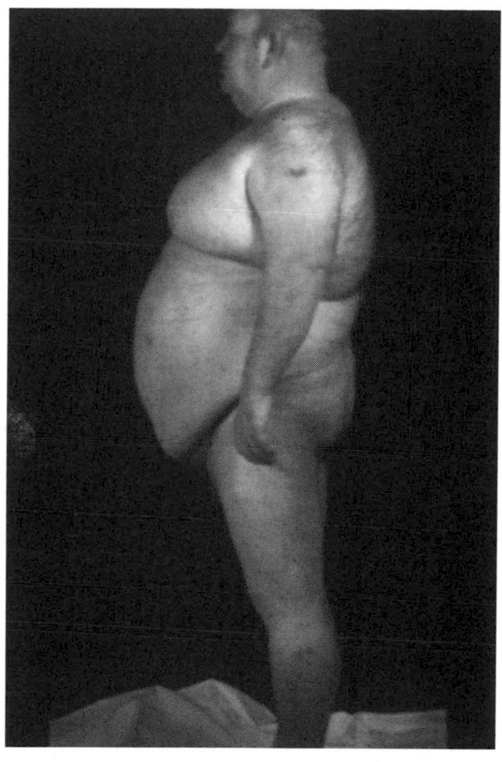

*One of the defining features of Cushing's disease, in which cortisol secretion occurs at higher rates than normal, is pronounced obesity, especially in the trunk region. Often, the arms and legs thin out as the trunk expands, as demonstrated in this individual. Because the primary cause of Cushing's disease is a tumor of the pituitary which secretes excess ACTH, surgical removal of the tumor usually restores cortisol levels to normal. After a period of time, the symptoms of Cushing's disease are reversed, and in many patients their body weight and fat distribution returns to normal. Another way in which Cushing's disease may occur, however, is by chronic administration of high doses of synthetic adrenal steroids like prednisone. When used to treat disorders like arthritis or to prevent rejection of transplants, these steroids can reach such high levels that Cushingoid symptoms arise. Fortunately, the effects are reversible once therapy is discontinued. (Copyright © Science VU/Visuals Unlimited. Used with permission.)*

which results from poor absorption of vitamin B12 through the intestines.

Today, it is estimated that there are at least 10,000 to 20,000 cases of Addison's disease in the United States. There are a variety of factors that contribute to the development of the disease, including tuberculosis (in which the bacterium invades the adrenal glands), autoimmune destruction of adrenal cells, and cancer of the adrenal. It is a slowly progressing disease, and because its symptoms are so generalized, it is often misdiagnosed as the flu or as psychiatric disorders. The similarity of the symptoms of Addison's disease to those of chronic fatigue syndrome (CFS), for example, have led to the hypothesis that some sufferers of CFS are actually in the early stages of Addison's disease. The treatment for Addison's disease is to replace the steroid hormones that are normally made by the adrenal glands, usually by oral administration of hormone tablets twice per day. Because the adrenal responds to stress by pumping out extra glucocorticoids, patients on steroid therapy are instructed to self-administer extra doses of the medication during times of illness, extreme anxiety, injury, or other stressful events. It is a difficult disease to manage, and must be carefully controlled each day of a patient's life.[5]

## Stone Age Stress: Real or Imagined?

So what can we conclude about the role of stress in our lives, now and in primitive times? If we imagine the types of physical stress our Paleolithic forebears must have encountered, it helps make our daily aggravations seem much less overwhelming. For example, prior to the advent of agriculture, the typical cave dweller would rarely, if ever, have the luxury of a steady, nutritious diet. On the contrary, malnutrition and starvation—with the vitamin and mineral deficiencies that go along with them—would have been extremely common. Sporadic dehydration from lack of clean or available water may have been as much a problem for primitive man as it is today for animals living in arid regions, like the plains of Africa. Hypothermia was a constant threat in the winter, especially in northern climes during the many ice ages. Injuries and infections that resulted from untreated minor wounds or parasite invasion would not only have been physiologically stressful but often lethal. Anthropological data suggest that our ancient ancestors suffered many of the same maladies that plague us today (arthritis, back problems, tooth decay, osteoporosis), but as stressful as those conditions are for modern man, they would have been far more stressful at a time when no medical treatment of any kind was available. Thus, it is not surprising that the life expectancy of ancient man was only a fraction of what it is today.

Likewise, our prehistoric ancestors did not need to negotiate city traffic and deal with short-tempered employers, but they certainly had their share of psychological stress. Not knowing when or if your next meal would come would have been (and continues to be for much of the world's population) a chronic source of anxiety. Each empty-handed trip back to the cave would have increased the tribe's fears for the next day. For that matter, obtaining a meal might have meant coping with the terror of chasing down an animal much faster and larger than oneself, using only sticks and stones as weapons. In fact, our early ancestors who first discovered the taste of meat were more than likely mere scavengers. A humbling thought, perhaps, but imagine chasing down and killing an animal the size of an aurochs (an extinct wild ox) or one with the speed of a gazelle with your bare hands.

A more subtle, but perhaps more terrifying source of stress for prehistoric man would have been his inability to exert control over his environment. Although an awareness of the cycles of nature and such physical principles as gravity probably existed in even our most primitive ancestors, an understanding of the forces of nature would have completely eluded them. Having no understanding of science meant having no sense of control over one's environment. Historical records of ancient societies indicate that people worried endlessly about celestial "beings" (sun gods, moon gods, and so on), and we know that until relatively recent times it was common for people to assign human traits to these deities. This would have implied that it was entirely feasible for, let's say, the sun god to feel angry or neglected one day and thus decide not to rise anymore. One can only guess, but it seems likely that the thought of perpetual darkness must have been as frightening a concept as could have been imagined by ancient peoples. Think of what it must have been like to go to sleep each night fretting that you may have failed to properly perform a certain ritual and that as a consequence your entire tribe or family might be doomed to darkness, chaos, and misery, that the rains may not come, or that winter will last forever.

We can see, therefore, that both from a physical and a psychological vantage point, our ancestors lived a much more stressful existence than we do today. The mechanisms that evolved to combat the deleterious effects of those stressors are still intact and usually serve us well. Yet we sometimes make things worse for ourselves. For example, today we have compulsive exercisers, people who can actually become addicted to strenuous exercise because this behavior imposes severe stress on a person's metabolism and results in the steady release of endorphin. Because this painkiller is similar to morphine in its addictive capabilities, it has been suggested that it produces a "runner's high." Extreme exercise also releases cortisol, which, though useful in maintaining circulatory and respiratory function, can lead to the stated problems, plus the suppression of fertility.[6] In yet another scenario, meeting a difficult deadline at work is a source of pressure, which, although it is not life-threatening, contributes to ill health by invoking an unnecessary release of stress hormones. The stress response thus evolved in animals to combat physical challenges to survival and was not meant to be activated for too long or too often. The downside of our wonderful forebrain, the seat of our consciousness, is that we have the ability to worry about things that aren't really threatening to our survival.

So, while there is no doubt that we are stressed in today's society, it is important to remember that all animals, including *Homo sapiens*, have always been confronted with innumerable types of stress. Moreover, from a historical perspective, it is clear that the incessant mantra we hear of ours being the Age of Stress makes no sense at all. Given the choice, who wouldn't prefer the aggravation of two working parents getting their kids off to school on time than the dread of being eaten in one's sleep by a lion?

CHAPTER TEN

# An Alternate Evolution

*We are passing from the animal into a higher form, and the drama of this planet is in its second act.*

—W. WINWOOD READE, AUTHOR, *THE MARTYRDOM OF MAN* (1872)

Have we reached the pinnacle of evolution, or is humankind evolving into a "higher form"? In the previous chapters we've seen many examples of superb evolutionary devices and defenses that demonstrate weak links in the human machine. So, the question is not whether we've already achieved physiological perfection but whether we ever will. I strongly suspect that the answer is "no," and not because of any doomsday predictions about the survival of our species.

A good way to understand our predicament is to compare the evolution of man with that of other animals, such as birds. Physiologically speaking, birds are in many ways superior to human beings. How and why did they get that way? Each year, as an exercise in thought, I put the following question to the students in my animal physiology classes: Suppose it was the dawn of time and God assigned to you the task of designing an animal capable of high-altitude flight. What features would you incorporate into such an animal?

The answers generally start, of course, with something along the lines of a lightweight, streamlined body shape with wings—lightweight so it can get into the air and streamlined so it can fly with less air resistance. To be light, a small body is necessary. On the other hand, massive pectoral muscles are needed to power the wings, but if the muscles are large, they're also going to be heavy, and that's going to impede flight. So the weight of other objects, like the bones, must be correspondingly lighter.

Because the animal needs to fly at high altitudes, where oxygen is scarce, a special pulmonary system is vital in order to capture extra oxygen with each breath—something like our own lungs, for example, only more efficient. In addition, telescopic eyes are needed to spot small, moving animals on the ground for food.

To meet the energy demands of frequent flight, a warm-blooded metabolism is also needed. To sustain the high body temperature, however, heat must be conserved in winter (especially at altitudes where temperatures can be extremely low). Covering the body surface with a good insulator like feathers is an ideal, lightweight way to conserve heat. In short, you'd need to build a bird.

Of course, the point of the exercise is to reinforce the idea that species tend to evolve characteristics that ideally suit their particular habitats. Try as you might, it is unlikely that you would be able to design an animal better suited to long distance, high-altitude flight than a bird. Indeed, birds are incredibly successful animals. One could argue that they are inferior to us, because we have a much larger, more complex brain, but in evolutionary terms, birds didn't evolve a brain like ours because they didn't need one. The fact that we did evolve such a brain suggests that it was necessary to overcome our many serious liabilities, or, conversely, we retained inferior sensory and physiological capabilities because we evolved a forebrain.

If we were to continue the exercise with other kinds of animals, we would be led to an inescapable conclusion: Compared to most species, human beings are in pretty sad shape! We're slow, earth-bound, weak, not terribly agile, and sensorially dull. Our sense of smell is woeful, our visual ability is almost useless at night, and we can hear sound only over a small range of frequencies and distances. Therefore, if we could go back in time and design a better person, what changes would we make? What qualities are we missing that would enhance our ability to survive? Naturally, we will assume that the perks of modern life are not available to help compensate for our deficiencies.

Let's start with the senses. If we were better suited to our environment, we would have larger eyes with more complex neural circuits in them to allow us to capture and process light images more effectively. To see better at night, we might have reflective crystals at the back of our eyes to help capture extra light, much like a cat or dog (think how a cat's eyes look when reflected in a car's headlights). We'd also profit from a better tracking ability to help us in hunting, but only if we had the speed to make use of it. So let's assume that our proto-human would be capable of reaching speeds closer to 50 or 60 miles per hour like a cheetah, rather than the 15 or so miles per hour a man can now reach, and still be able to focus on the world around him. This would require a redistribution of muscle mass to power the legs and a respiratory system capable of keeping up with the increased oxygen demand. These changes alone would make for a far superior person better suited to exploit a primitive environment.

Of all the senses, the sense of smell probably gives an animal more information about its environment than any other. This is partly because signals from odor receptors in the nose are directly linked to the brain centers that control learning and memory, which is why smells elicit such powerful memories. But our sense of smell is the weakest of all our senses. Other animals, even insects, can detect odor molecules that are many thousands of times less concentrated than we need them to be. Although an animal with great vision—like the mythical proto-human we are designing—would be less reliant on olfactory capabilities, any increase in sensitivity could mean the difference between survival and becoming a meal for a lurking predator.

Our sense of hearing could also be improved. Before the discovery of electricity, we could only communicate immediately as far as we could yell. If we could detect the low frequencies that an elephant hears, we could talk to each other over a distance of several miles! Of course, to pick up those sound waves from so far away we'd need pretty big ears, much bigger than our present ones. To make sure we could detect the direction of sound, we'd need to space our ears as widely apart as possible, which can only be done by having an enormous head or long ears that project away from the head. It would be even better if we were able to individually rotate our ears from front to back and up and down—much like owls and bats—to help us identify the source of a sound. Compared to the look that we are used to, we might look pretty comical, but no doubt over time we'd consider tall, floppy, movable ears attractive.

We might also be better off if we more closely resembled our primate cousins and had a thick coating of hair or fur to help protect us from the elements and retain heat. In order to exploit many different environments, however, we'd need to lose that fur if we migrated towards hot climates. In other words, we'd need to be able to shed our fur in summer and regrow it in winter. And to help us blend with our surroundings and avoid detection, it would be useful to have white fur in the winter and brownish-green fur in summer.

There are countless numbers of additional, subtler changes we could make. How well our kidneys function, how powerfully our heart pumps, how well enzymes in the body function under extreme conditions, and so on, could all be improved. It's humbling to realize how little of our anatomy and physiology is anywhere near perfection.

The question thus arises, why *didn't* we develop into a more impressive animal than what we are today? Almost certainly it's because we didn't need to, because we developed something no other animal has yet matched—the human forebrain. The neocortex of even the smartest animals—chimps, dolphins, and (yes) pigs—doesn't even remotely compare to ours. Generally, if an animal has a really superior feature, it tends not to develop other features to quite such an extreme. As we've seen, birds have extraordinary vision, but a dismal sense of smell. In our case, the evolution of a large, complex brain was such a momentous step forward that virtually all the other features needed for survival in the wild failed to develop to levels seen in our less intellectually gifted mammalian relatives. Thus, we never developed much speed because we learned to hunt and ambush prey. We aren't strong but we created weapons. We didn't need a layer of fur because we learned to use animal hides to cover ourselves. So, maybe we are already nearly perfect after all—for our purposes. As the noted anthropologist Desmond Morris once pointed out, cats may seem dumb to us but they are very, very good at being cats.

What will the next million years of human evolution bring? That's hard to say, because it's likely that we ourselves will continue to direct its course with advances in medicine and genetic engineering. But in the complete absence of scientific and medical interventions, how might we look in some distant future? For example, would our appendix finally disappear altogether? Would our spine finally adjust to our upright posture, alleviating the common backache? Would we finally stop growing wisdom teeth? Would the birth canal finally widen to more easily accommodate the head of an eight-pound baby? Would our forebrains continue to develop, possibly imparting new intellectual powers? It's perfectly reasonable to assume such changes might occur and not necessarily over a timescale of millions of years. In the case of the *Anolis* lizards in the Bahamas, just such a thing has happened. Small populations of these reptiles were seeded onto a number of tiny islands in the late 1970's. In the 20 years since then (equivalent to 14 generations), lizards seeded on islands where the predominant vegetation was narrow branches and twigs have grown shorter, with stronger, stumpier legs, which has better adapted them for life in their new environment. In other words, these populations of lizards appear to have *evolved* into a new form in the course of just 14 generations. Similarly, certain species of guppies living in tanks without predators grow colorful, large tail fins, but when predators are around, these guppies stand out and are the first to be eaten. Their less colorful kin evolve into a new form very rapidly. The best case for relatively quick evolutionary changes in humans appears to be the Sherpas, who live on the high southern slopes of the Himalayas and provide support for mountain climbers. Over the course of millenia, they have apparently evolved cardiopulmonary features to make them better suited to life in the Himalyas. These features are truly genetic changes, as they are passed to their offspring and do not disappear even when a Sherpa moves to sea level.

What's ultimately relevant for any species, however, is how successful it is in adapting to its environment and continuing the species. If you consider which animal best fits this description, it's surely the insect! Insects have been around for ages, make up roughly half of the million or so known animal species, and will undoubtedly outlast humankind. (Maybe we should have incorporated some insect-like qualities into our proto-human, like the sci-fi flick *The Fly*.)

Although no one can predict the future or which species will manage to hang on as the coming millenium comes to a close, one thing is certain. Those animals alive to welcome in the year 3000 will be those that were best able to cope with the fundamental challenges to the survival of all organisms, big and small—food, water, salt, energy, oxygen, and temperature. As long as the Earth supports carbon-based life-forms like ourselves, these needs will never change. Every species—no matter how they evolve over the next millenia—will need to deal with these challenges just as they do today and have for countless millenia before.

EPILOGUE

# Doing Physiology

When William Harvey concluded his letter to the president of the Royal College of Physicians, he might have rewritten the last line, "Think kindly of your anatomist," as "Have pity on your anatomist (or physiologist)." Physiology is a difficult science to study experimentally. It takes a lot of imagination, creativity, perseverance, and a bit of luck. It's possibly the only branch of modern science in which a combined education in chemistry, physics, mathematics, statistics, biochemistry, molecular biology, engineering, and of course good old-fashioned biology are absolute prerequisites. An electrical engineer can do his or her job perfectly well without the slightest inkling of how the human body works, but a physiologist interested in how brain cells communicate must be fully versed in the principles of current, resistance, capacitance, feedback, and electrical decay, all of which are fundamental principles of the physics of electricity. Likewise, a chemist interested in synthesizing new commercially useful chemicals need know nothing about how the heart works or how a hormone activates a target cell, but a physiologist that studies these and other processes must be well versed in the chemical nature of the relevant molecules, such as hormones or neurotransmitters that speed up the heart. To study how the human (and animal) body functions, one needs to understand a wide range of disciplines.

Although this is a tall order, it's ultimately very rewarding. The breadth of training of physiologists makes them extremely valuable in tying together the reductionist with the organismal or population sciences. It also makes for an intellectually stimulating life. My colleagues in other disciplines will no doubt disagree, but I would find the life of a "pure" molecular biologist personally unsatisfying.

It may be true that in today's scientific world researchers need to specialize, but it shouldn't be to the point of being unable to appreciate the significance and value of different research approaches. It sometimes seems that some of this appreciation has been lost. In part, this stems from the fact that the training of young molecular biologists today can be rather narrow. The result can be a brilliant researcher who has difficulty recognizing the validity, and need, for alternative approaches, like population biology. Heaven help us if we ever channel our energies completely to a reductionist training program. Without people trained to see the "big picture"—how a gene product may affect not just a cell but a whole organism or even a whole population of different animals, for example—we'd be in serious jeopardy.

## Asking Questions and Getting Funded

Like all scientists, physiologists begin with a question, or a hypothesis. A research plan is conceived to answer the hypothesis and experiments are conducted. The results either support or refute the hypothesis. Of course, part of being a good scientist is being able to acknowledge the importance of a refuted hypothesis. Oftentimes, the most exciting scientific discoveries arise from rejected hypotheses, that is, we think an experiment might turn out one way, but it turns out the opposite way—and the opposite way is more interesting than the predicted one. It is the scientist who tries desperately to fit the observed data into a preconceived idea of how the results should look who ends up a failure.

But how are these experiments funded? Primarily through grants from government agencies, philanthropic societies, and industries. Pharmaceutical corporations have also become major funding organizations in recent years, as our ability to churn out better and more sophisticated drugs for treating disease has exponentially improved with new technologies. A major difference between taxpayer-supported research and privately funded research is that the latter can place restrictions on what gets published and when. A major pharmaceutical firm, in competition with others, has the right to tell its scientists not to publicize any of its important findings until the company decides it is ready. In the case of a potentially lifesaving new drug, this raises provocative issues of ethical conduct that society is just now beginning to address. Government funded research, on the other hand, is usually published as rapidly as possible without constraints dictated by financial self-interest or greed.

If a scientist like myself wants to test an hypothesis about something, I must first apply to one or more of these funding agencies for money to buy supplies and equipment, and then pay the salaries of technical staffs as well as part of my own salary. To be successful, the application must be approved by an extremely rigorous peer review system. Virtually any flaw in logic in an application is grounds for rejecting the application. In fact, in the past 10 years the funding process has become so competitive that many young (and some older) scientists have been forced to abandon their years of training and seek other types of employment. This amounts to a tremendous waste of time, effort, training, money, and expertise, and even those who have been successful in getting grants find that for each application that gets approved, anywhere from two to ten are rejected.

A typical physiologist works with a grant-supported budget of approximately $100,000 to $200,000 per year. Where does all that money go, apart from salaries for staff, students, and technicians? Most of it goes toward the purchase of supplies and equipment. For example, a gamma counter, which is used to measure the amount of radioactivity in a sample, costs about $30,000. A centrifuge, used to separate blood and tissue culture samples into multiple components, runs anywhere from $1,500 to $50,000. Assay kits used to measure the amount of a particular hormone in the blood may run as high as $300 per kit, and a lab like mine may go through 100 such kits each year. Suffice it to say, it is a rare occasion when anyone has a surplus left in their budgets at the end of the fiscal year.

Once a grant is awarded, the research gets underway, and eventually the results are written up and published in a peer-reviewed scientific journal. By this time, the grant-writing stage has already come full circle, as one grant ends and another is applied for, each typically lasting one to four years.

## The Field or the Laboratory?

Once the hypotheses are formulated, it is time to get down to the business of actually doing the experiments. It is difficult to design experiments with live animals that minimize uncontrolled variables and still permit a logical interpretation of the results. Scientists like to tell the old joke about a researcher who wants to determine the way in which frogs hear and proposes the hypothesis that the hearing organs are located in the legs. He first has to train the frog to respond in some way to a controlled sound, so that he can figure out when and if the frog is actually hearing something. Eventually, the scientist trains a frog to jump whenever he claps his hands. He then removes the frog's legs and claps again. The frog doesn't jump. "Aha," says the scientist, "I was correct. Without his legs the frog cannot hear."

A silly joke, perhaps, but the point is valid. When working with animals, it is not always easy to be sure you've controlled for every variable in an experiment. It can also be quite difficult to mimic conditions of the "wild" in a laboratory. As you'll recall from Chapter 5, the remarkable bar-headed goose migrates each year between India and Tibet, flying over the Himalayas, where the oxygen pressure is extraordinarily low. This is a feat unmatched by any other air-breathing vertebrate in the world. In order to understand how these geese cope with the low oxygen pressure under such conditions (perhaps because you hoped to apply this knowledge to the human condition, for example, to assist people with lung disorders), you would need to study it and take various measurements while it is flying. Since this is impossible, the only alternative is to bring the birds back to a laboratory and study them there.

The problem with such an arrangement, of course, is that the bird has been placed in a completely unnatural setting. The air in the room is at sea level pressure, which means more oxygen pressure than is found over the mountains. The temperature of the air is also much warmer than it would be at that altitude, and cold air can have numerous effects on bronchiole function (just ask any asthmatic). Finally, what you're really interested in is how the bird manages not only to exist but to exercise (fly) under those conditions, because human mountain climbers are nearly incapacitated at that elevation. Thus, the arrangement shown in the accompanying figure was devised. The bird is trained to run on a treadmill (!) in order to simulate the exercise performed in flight, and it breathes through a nose cone an air mixture that is composed to mimic the air atop the Himalayas. Various blood sampling catheters are placed in the veins and arteries of the bird in order to monitor blood oxygen and blood pressure during the test. Quite a setup for a single goose!

In some ways, experiments on human beings are easier to perform, because you can explain to the subject exactly what's required of him or her, but such experiments are bound to be limited. You obviously cannot perform brain surgery on someone, for example, just to

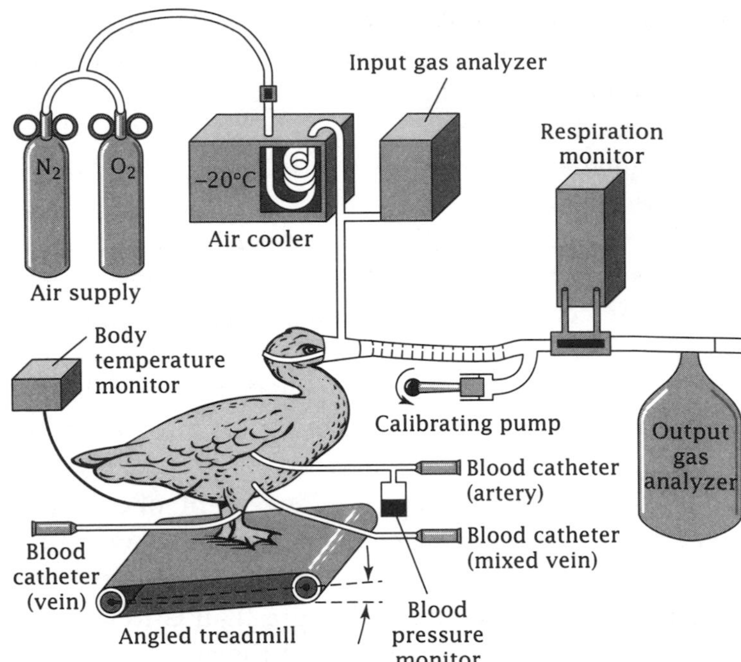


*Redrawn from M. R. Fedde et al., "Cardiopulmonary function in exercising bar-headed geese during normoxia and hypoxia,"* Respiration Physiology, *volume 77, pp. 239–262, 1989, with kind permission from Elsevier Science–NL, Sara Burgerhartstraat 25, 1055 KV, Amsterdam, The Netherlands.*

determine how a part of the brain works. There is also the danger of unforeseen problems in human experiments. Let me give you an example from my own personal experience. Many of the subjects of clinical physiological studies are graduate students that are willing to do almost anything for a buck. I was once one of those and volunteered to test the effectiveness of a new drug for preventing the onset of the symptoms of very mild asthma (which I've had since infancy). In the very first trial, I was given an inhaled dose of atropine, a drug that was supposed to prevent any symptoms of bronchoconstriction when I was later exposed to an allergen. The drug did indeed do what it was supposed to do, but, to my horror, I developed several unpleasant side effects that persisted for hours. It turned out I was mistakenly given too high a dose (I was the first subject in the study, truly a guinea pig among guinea pigs), and the effects were dilation of the pupils (on a sunny day!), swelling of the fingertips, racing of the heart, cold sweats, and, most disturbing of all, an inability to urinate for more than 12 hours. (I did not continue the study, by the way, but insisted on receiving payment for the full 10 sessions). Those studies, however, did lay the groundwork for what has become part of the arsenal against childhood asthma. Atrovent, a derivative of atropine, is a commonly prescribed aerosol inhaler that helps keep airways open in children with asthma. So I suppose I had a small, if ignominious, part in medical history—at least in helping them get the dose right.

Many physiologists eschew animal studies entirely, preferring to study isolated cells or tissues in a culture medium. These types of experiments are much easier to control, are usually more reproducible, and are cheaper. They are also extremely useful for understanding the cellular mechanisms that drive physiological events, like the contractions of the heart. But this approach is also limited by the inescapable fact that heart muscle cells (or any other), enzymatically shorn apart from each other and placed in a culture dish, may behave quite differently under such artificial conditions than they would inside the living body. Thus, results obtained using tissue culture methods must eventually be reproduced in a living animal before anyone would be confident enough to extrapolate the results to the human condition.

And so it goes. Physiology has come a long way since the dark days when scientists were also grave robbers by night, in order to obtain specimens for study. It's a very, very old science that has traversed hills and valleys in popularity over the millenia. Nonetheless, it remains the cornerstone of modern medicine and will continue to do so into the next millenium.

# NOTES

## CHAPTER 1 Different Species, Same Problems

**1.** The actual derivation of the word "physiology" comes from the Greek *physis* (nature) and *logos* (study). Whether or not the *physiologoi* were more folklore than real, the science of physiology does begin with the ancient Greeks—among them Aristotle (384–322 B.C.) and Hippocrates (ca. 460–377 B.C.). Most of these early philosopher-scientists used the art of dissection to analyze the appearance and interrelationships between internal organs. Aristotle's contributions were many, including being the first to try to classify all living species of animals (he managed to round up about 500, which is pretty good, but about 1 million fewer than we've now classified). Hippocrates, after whom the Hippocratic oath taken by physicians is named (though it's not certain he was the oath's author), was a physician who performed many animal dissections to determine the nature of disease. Around 400 B.C., he came up with the idea that disease was not due to the wrath of the gods, as traditionally believed, but was caused by some disturbance of the "humors" of the body. Fever, he suggested, was the body's attempt to rid itself of illness. This can be considered the forerunner of the modern idea of homeostasis and was quite a remarkable brainstorm. For those interested in the Hippocratic oath, I've reproduced an amalgam of various translations of it here. New doctors recite an updated version of the oath to this day. Note that even during Hippocrates' time, abortion and assisted suicide were issues of deep concern to physicians.

> *I swear by Apollo Physician, by Aesculapius [or Asclepius, the Greek god of medicine], Hygeia, by Heal-All, and by all the gods and goddesses, that I will carry out this oath according to my ability and judgment: To regard my teacher in this art as equal to my parents; to share my*

*substance with him, and when he is in need of money to share mine with him; to consider his offspring equal to my brothers, to teach them this art if they wish to learn it, without fee or stipulation; and to impart precept, oral instruction, and all the other learning, to my own sons, to the sons of my teacher, and to pupils who have signed the indenture and sworn obedience to the law of medicine, but to none other. I will use treatment to help the sick according to my ability and judgment, but I will never use it to injure or wrong them. I will not give poison to anyone even if asked to do so, nor will I suggest such a plan [in other words, physician-assisted suicide]. Nor will I give a pessary to a woman to cause abortion. With purity and holiness I will spend my life and practice my art. I will not use the knife on sufferers from stone, but will instead leave this to be done by those who practice this work. Into whatever houses I enter, I will do so to help the sick, keeping myself free from all intentional wrongdoing and harm, especially from fornication with woman or man, be they freemen or slave. Whatever in the course of practice (or outside my practice) I see or hear that ought never to be spoken in public, I will not divulge, but will instead consider such things to be secrets. If I keep this oath, and not break it, may I enjoy honor in my life and art, among all men for all time. If, however, I violate this oath, may the opposite befall me.*

The first truly influential physiologist was the physician Galen, born to Greek parents in present-day Turkey about 130 A.D. Galen was the first person, as far as we know, to systematically study how the parts of the body function by performing experiments on animals. Legend has it that he chose a career in medicine as a result of a dream, in which Asclepius appeared and told Galen to tell his father of the future that was planned for him. This was a fortunate happenstance for mankind and saved Galen from a life as a statesman or architect, careers his father preferred for him. More than likely, however, it was a ploy by the young Galen to convince his father to let him pursue a career as a physician.

Galen held the position of physician to the gladiators, which is undoubtedly where he learned firsthand how the parts of the body function in living (and dying!) bodies. He was the first to describe many of the nerves and muscles of the body and to assign them specific activities, such as respiration, and was an early proponent of the importance of proper diet in maintaining health and fitness. He also came up with the idea that the muscles of the heart had an intrinsic ability to beat even if separated from the rest of the body (although this was certainly already known, because anyone who has seen a recently killed and dismembered animal or person would notice that the heart can continue to beat erratically for a brief time). He also provided a careful survey of anatomy by performing dissections on the Barbary ape. Galen was the first to demonstrate that the arteries contained blood. Before Galen, it was believed that the veins but not the arteries contained blood, because anatomists dissected dead animals, in which blood pressure was no longer present; upon opening an animal, the blood drained out and the arteries appeared empty. Galen, however, used live animals in his experiments.

Galen also came up with the idea that there were two sources of pressurized blood in the body. The liver, he believed, converted food into blood, then pumped it to the heart and elsewhere (in what way, he hadn't any idea). This "liver blood," he reasoned, contained the ingredients derived from food that somehow nourished the body. Liver blood was said to be consumed by the body, and additional food was needed to replenish it. The heart, he thought, pumped out another type of blood that contained some essential "spirit" essence that imparted a life force to the body. The two sides of the heart, he believed, communicated through tiny holes in the septum that separates the chambers. This view would be accepted for nearly 1,500 years, although the occasional skeptic challenged the role of the liver in blood production. Nonetheless, most people from Galen's time on falsely believed that blood was continually produced and consumed by the body. The idea of a fixed quantity of blood that was recirculated over and over again was a vague notion that bounced around from time to time but wasn't proven until the seventeenth century (for more, see Chapter 6).

Although this latter idea of Galen's was completely wrong, and his name has become somewhat forgotten in modern times, he was nevertheless the first great experimental and theoretical physiologist and one of history's great thinkers. He died about 200 A.D., but his influence extended for more than a thousand years. His greatest surviving treatise is *Methodus Medendo*, which summed up the whole of medical knowledge of his day and was the bible for physicians well into the next millennium. In fact, words were coined to describe Galen's practices and the physicians who followed them (Galenism, Galenists), and Galenical medicine used the herbal dietary practices promoted by Galen. It might be argued, however, that Galen's giant stature among physicians actually delayed scientific progress, because few people had the wherewithal (or the courage) to challenge his ideas.

From Galen's time through the Middle Ages, most of the great discoveries in physiology were made by the Chinese, Italians, and Arabs. For example, by the seventh century, Chinese physicians were aware of the classic symptoms associated with diabetes, such as extreme thirst and a sugary urine, but it was not until the twentieth century that Western scientists were able to deduce that these symptoms were due to a problem with the pancreas. From the Renaissance to the twentieth century, the British, French, Italians, Flemish, and Germans led the way in medical advances, while American physiologists made their greatest contributions in the twentieth century.

For a wonderful but concise timeline of major historical advances in the life sciences, and the physical sciences as well, see A. Hellmans and B. Bunch, *The Timetables of*

*Science: A Chronology of the Most Important People and Events in the History of Science* (Simon and Schuster, 1988).

**2.** Although the idea of homeostasis began with Hippocrates, it wasn't fully elaborated until the French physician Claude Bernard (1813–1878) came up with the idea that the *milieu interieur* (internal environment) of the body was normally maintained in a constant condition. In other words, the fluids of the body can adjust their composition in ways to counteract changes imposed by the environment, and we keep a warm body temperature even when it's very cold outside. This was in many ways a major step forward in understanding how the body works but was not completely accurate, primarily because of Bernard's use of the word "constant." In fact, most physiological functions operate around a fairly stable baseline, but fluctuations above and below the baseline occur all the time. The daily swings in body temperature are a good example. The word "homeostasis" was not coined until some years later, by Walter Cannon of Harvard University, one of the great intellects of the twentieth century. For more on Cannon, see Chapter 9.

## CHAPTER 2 1,000 Cheeseburgers for Lunch

**1.** We all know that oxygen is needed to burn fuel, but does that mean we have tiny little fires flaring up in all our cells? It's better to think of what goes on inside our bodies as controlled combustion, which generates heat but not fire. Nevertheless, the heat serves to keep us warm, and when metabolism is higher than normal, we can overheat, and when it is lower than usual (as it is in the middle of the night), we feel cold. What does the oxygen actually do in these reactions? The glucose and fatty acids that enter cells are acted on by enzymes, which convert the molecules into smaller ones, bit by bit. At many of these step-wise reactions, bits of energy are released and stored in the form of a molecule called ATP (adenosine triphosphate). When needed, ATP can release its stored energy to drive any of a thousand cellular reactions, like those that power the contraction of muscles. Without oxygen, however, this process is very inefficient. At the final step in the breakdown of glucose, for example, there is a stray electron that has been jumping from one enzyme to another (creating ATP in the process). If this electron isn't soaked up by something at the last step, no more electrons will be able to follow it down the energy-making slide, and the reactions will stop. When this happens, a condition known as anaerobic respiration (making ATP without oxygen) begins. Fortunately, the chemical nature of oxygen is such that, when present, it will act like a sponge to pick up the stray electrons and allow the reactions to continue, whereby another molecule of glucose gets broken down and more ATP is made.

Although the amount of ATP formed by anaerobic respiration is trivial compared to that formed when oxygen is present, it is very useful on occasion, especially in animals that are under water for long periods of time. One problem with anaerobic respiration is that it generates a dangerous waste product known as lactic acid, which gives our muscles that leaden feeling when we exercise too strenuously.

**2.** The notion of how the anatomy and physiology of different species change with body size, known as scaling, is a fascinating and favorite preoccupation of comparative physiologists. Some of it is common sense, such as the heart being a constant percentage of body weight in any healthy mammal. If I weigh 180 pounds and you weigh 120 pounds, your heart is about two-thirds the size of mine, assuming neither of us is significantly overweight. But the business of metabolic rate not being strictly proportional to body mass can be confusing. (When you graph metabolism versus body size, you don't end up with a straight line.)

The exact mathematical relationship between body mass and metabolic rate looks like this (omitting arithmetic constants for the sake of simplicity): metabolic rate = body mass$^{0.75}$. Thus, as an animal gets twice as big (a factor of 2), its metabolism only increases by a factor of about 1.7. Another way of looking at this is to say that as animals get smaller, their metabolisms do not get proportionally smaller. An animal half the size of another doesn't have a metabolism half as much but rather a bit more than that, about 60 percent of the larger animal's metabolism.

We can begin to see why this problem has confused so many people for so many years. In the late nineteenth century, the German physiologist Max Rubner proposed the Surface Rule, which, briefly, stated that as an animal gets larger, its mass (or volume) increases to the third (cubed) power, but its surface area increases only to the second (squared) power. Therefore, if you increase the linear dimensions of an animal by a factor of 2, the animal's mass increases by $2^3$, or eightfold, while its surface area increases only by $2^2$, or fourfold. Surface area, then, varies to the two-thirds power (0.67) of body mass. This in itself was pretty elementary arithmetic and could apply as easily to a bowling ball as to an animal, but Rubner then went on to postulate that metabolic rate should be exactly proportional to surface area. Why? Because heat loss of any object is exactly proportional to surface area, and metabolism is what is responsible for generating heat. Thus, tiny animals lose heat more easily because of their *relatively* greater surface area and must have a disproportionately high metabolism to compensate. A beautiful hypothesis, except that when you do the experiments, the exponents don't add up.

About 50 years after Rubner formulated his hypothesis, Max Kleiber came up with the 0.75 exponent described above. That is, metabolism varies with the 0.75th power of body mass, *not* the 0.67 power as predicted by

Rubner. Close enough, you might say and not worth quibbling over. But these numbers are an exact science, and although they are close enough to at least consider that the Surface Rule hypothesis is partly true, they are far enough apart to prove that the hypothesis is not the only answer.

Perhaps the best explanation of these and numerous other fascinating relationships is found in K. Schmidt-Nielsen, *Scaling: Why Is Animal Size so Important* (Cambridge University Press, 1986). In that book you'll learn that in reality surface area and body mass don't change with the 0.67 power as predicted but rather to the 0.63 power (even further away from the experimentally observed 0.75). This is because the shape and contours of an animal don't change equally in all proportions as you go from smaller to larger species. An elephant, for example, has additional mass in its legs to support the weight of its enormous body. A mouse, on the other hand, has very little of its mass in its tiny legs.

**3.** As body size doubles, lung size doubles. (The exponent in the equation would be exactly 1.) Keep in mind that we are comparing animal species with each other. I'm not saying that if a person who weighs 100 pounds puts on 100 additional pounds by overeating that his or her lung size doubles. All of these comparisons assume that we are comparing adult, lean, healthy, "typical" members of one species to those of another.

**4.** Although this chapter focuses on the need for food to drive metabolism and the consequent need for oxygen to convert food to energy, it's important to bear in mind that waste products are generated by these activities. These can be quite toxic, especially if allowed to accumulate in the blood at high concentrations. For example, as discussed in greater detail in Chapter 5, it is just as important to get rid of $CO_2$, the major waste product of burning fuel in the body, as it is to take in oxygen. $CO_2$ is a very deadly compound if it reaches levels in the blood that are only 50 percent or so above normal. In fact, it is so dangerous that it is $CO_2$, and not oxygen, that primarily drives the brain to keep us breathing regularly, even when we're asleep.

## CHAPTER 3 Too Much to Eat!

**1.** Let's define some technical terms. A nucleus in the brain is defined as a cluster of brain cells in a compact area that serve similar functions. Some nuclei may be quite small—only a few hundred brain cells—while others may number in the thousands or more. They can be detected with certain dyes that stain the cells a color, usually brown. The individual cells within nuclei are called neurons. Many people use the terms "neuron" and "nerve" interchangeably, but technically they mean two different things. A nerve is a collection of neurons whose processes, or axons, all travel along the same path, either within the brain or out the spinal cord. Some neurons within a nerve may send their axons out at certain branch points to stimulate muscles or other structures. Other neurons may continue to travel with the nerve for greater distances until they finally reach their destination.

**2.** The purpose of all those inputs and cross-wiring networks is probably to ensure a stable system with lots of built-in checks and balances. For example, we know that one of the most widespread neurotransmitters in the brain is a small molecule called GABA (gamma aminobutyric acid). GABA is an inhibitory transmitter, and every time GABA comes in contact with a neuron, it makes the neuron quiet down or become less "jumpy." A brain deficient in GABA would be hyperexcitable, somewhat similar to the way that parts of the brain become hyperexcited during epileptic attacks. In essence, the brain is like a car being driven with both the accelerator and brake on at the same time. Removing the brake (GABA) causes the brain to accelerate.

**3.** For a discussion of CRH, see Chapter 9. CRH is the major stress hormone of the brain and is responsible for numerous actions in addition to its presumed control of feeding and energy balance. Without CRH, it is generally believed, mammals could not survive unless they were treated with steroid hormones derived from the adrenal glands. That's because CRH initiates a cascade of events that culminates in the production of cortisol from the adrenals. It is the cortisol, actually, that is an absolute requirement for life. Most people are familiar with cortisol as an over-the-counter skin cream known in pharmaceutical circles as hydrocortisone.

**4.** What exactly does insulin do? It sounds disarmingly simple but is actually rather complex. As molecules of sugar (known as glucose) circulate in the blood after a meal, they need to leave the blood and get inside the cells of the body. Within the cells is the machinery needed to convert glucose into energy units, called ATP. Very active cells, like those in our muscles, need more ATP and thus use up a large percentage of the glucose in the blood. Here's where insulin comes in. Molecules of glucose are too big to move across the membranes that cover and protect cells. Insulin has the ability to make the membranes leaky to glucose, and the more insulin, the more easily glucose can move across the membrane barrier and into the cell interior where it's needed. Without insulin, as in insulin-dependent diabetes, glucose cannot get across the membranes, and the cells begin to starve and die.

The first warning signs of the development of diabetes are usually dehydration, constant thirst, constant need to urinate, and sometimes dizziness and blurred vision. All of these symptoms are related to the buildup of glucose in the bloodstream. Since glucose cannot enter the cells, it remains in the blood. What's more, the brain and liver recognize that the body is starving and begin to break down fat and muscle into smaller units that can be made into yet more glucose. It's a vicious cycle. The glucose gets so high in the blood that it "spills" into the urine. This is a classic marker of the disease and is very easy to test for. When you go for a routine physical and give a urine specimen, a specially treated dipstick is placed into the specimen and turns color if glucose is present.

As glucose leaves the body in urine, it draws water with it, and diabetics thus become very dehydrated and feel terribly thirsty. I once had a singular experience while taking a cab ride across town to Boston University's medical school to give a seminar on this very topic. During the ride, the cabby and I got to talking, and he told me that he'd been experiencing strange changes in his daily "rhythms." Now, normally, this is where I'd ask to be let out of the cab, but something about his demeanor interested me. He told me he'd been getting up four or five times during the night to use the bathroom, something he never had to do before. He also found that he was drinking gallons of water a day. At this point, I was looking for the hidden camera, as I could hardly believe that this stranger was telling me that he was developing the symptoms of non-insulin-dependent diabetes while taking me to my seminar on the topic (unbeknownst to him, I might add). I asked him his age (50, right around the age when the disease commonly develops) and if he had put on weight recently (yes, about 25 pounds over the past few years). He asked me if I had ever heard of symptoms like his. By this time, we had reached the main hospital at BU, and as we pulled in I told him to park his cab and go directly to the diabetes unit in the hospital. He looked at me incredulously, and I don't know if he took my advice, but I noticed him lingering there in the cab as I made my way inside. I hope for his sake he listened, however, because untreated diabetes can progress frighteningly fast to serious illness and coma. Obesity is one of the leading factors that contributes to this type of diabetes, and oftentimes exercise and weight loss are sufficient to restore normal sensitivity to insulin.

## CHAPTER 4 Getting Enough to Drink (Water, that Is)

**1.** Although it may seem hard to believe, too much water can also be harmful. In the 1970's and 1980's a fad diet became quite popular in which people drank excessively large amounts of water each day. The diet was actually more harmful than helpful, at least to some unfortunate souls. Drinking too much water leads to a rare condition known as "water intoxication," in which the osmotic balance of the blood, interstitial space, and intracellular fluid is completely out of kilter, but in the opposite way from the case described in the text.

**2.** Animals that live in the desert, such as the kangaroo rat, recapture metabolic water by exhaling air over a coolant system built into their nasal passages. As the warm air leaves the lungs, and passes over this cooler area in the nose, the water vapor condenses and is reabsorbed. Thus, the air exhaled by a kangaroo rat is much drier than that of other animals.

**3.** Lungfish are found in only a few parts of the world, notably Australia and parts of Africa and South America. They are intriguing because they possess both a lung that can breathe air and gills that can breathe water. Thus, when water is stagnant and the oxygen content of the stream or river in which lungfish live drops, they can emerge from the water and gulp air for a while. They also have four strong fins that function like primitive legs on land. Because of these features, they are considered a link between fish and modern-day amphibians. Lungs have evolved separately in some invertebrates, too. Some arachnids, like spiders, have a primitive type of lung called a book lung. It doesn't really function much like our lungs but is a lung nonetheless.

**4.** The relationship between volume and pressure is a basic principle of chemistry and crops up over and over again in various settings throughout biology. It is based on the physical principle known as Boyle's Law, which states that when a fluid- or gas-filled chamber increases its volume, the pressure of the water or gas in that chamber decreases. As one goes up, the other goes down. On the flip side, if a chamber is compressed, the pressure within the chamber increases (think of a piston and cylinder in your car engine). This law also explains how the lungs fill with air.

You might imagine that breathing water over the gills could fill the gills with all kinds of muck. Ocean water is loaded with tiny fragments of dead or decaying organisms, pollutants, plankton, and other junk. This could accumulate in the gills and make it harder to get oxygen. To solve this problem, a fish does the same thing we do if airborne garbage gets in our airways; it coughs. Yes, fish cough, and they do so by suddenly reversing the direction of water flowing within the gill and mouth chambers. In short, they cough the water back out into the sea with a strong muscular effort, thus dislodging the debris in the gills.

**5.** Under a microscope, the gills are seen to be made up of wafer-like structures called lamellae. Inside these extremely thin structures are enormous numbers of small capillaries. Oxygen moves from the ocean water, through the cellular membranes of the lamellae, and into the capillaries, where it is carted off to supply the needs of the fish's body. The lamellae are so thin and fragile that they are unable to support their own weight in air, which is why fish die when taken from the water. The lamellae collapse and stick together, and the surface area available for oxygen exchange becomes drastically reduced, causing the fish to suffocate.

**6.** Fish breathe water at a high rate, which takes energy. If the amount of oxygen dissolved in a tank of water is reduced, fish can be seen opening and closing their mouths and opercula at a faster rate (they also do this when sick). The same thing happens to us if we breathe air in which the oxygen has been partly depleted. When we climb a mountain, for example, the oxygen pressure becomes lower, and we automatically respond by breathing faster and deeper (hyperventilating). This topic is explored in detail in Chapter 5.

**7.** The first signs of a drop in blood volume (called hypovolemia) are headaches and a tendency to faint or get light-headed when standing up suddenly. Fortunately, it's easily corrected with an infusion of fluid. Loss of water from the body also causes the blood to thicken. When the water content falls, the concentration of red blood cells therefore rises (the total number of cells stays the same, but they are packed into a smaller volume). The percentage of blood that is comprised of red blood cells is called the hematocrit.

**8.** The most common of these sports drinks is Gatorade, and there is a wonderful story behind its invention. The idea came from Robert Cade, a physiologist at the University of Florida who in the late 1960's was the football team's physician (among other responsibilities). As fans of college football know, "Gators" is the nickname of the University of Florida team at Gainesville. Football practice generally starts each year in late summer, and in Gainesville that means temperatures in at least the 90s every day. Strenuous exercise, especially in football gear, can cause dehydration under the best of conditions, but in 90-degree heat it can be very serious (even fatal). Thus, you might imagine that the Gators should be drinking plenty of water to keep from becoming too dehydrated, but that is not the best solution (pun intended!). Ideally, to avoid dehydration, you need to drink something that exactly replaces what was lost by perspiration. That way, you not only maintain body water levels but also return your salt concentrations to normal (perspiration contains both water and salts, especially sodium and potassium).

Dr. Cade had been asked by one of the football players why players don't need to urinate during the games, and even for a while afterwards—a phenomenon familiar to anyone who becomes dehydrated. The answer is because the kidneys temporarily power down to a lower operating level in order to preserve water. For the football players, however, dehydration meant that their performance started to wane in the second half of the game. So, Dr. Cade came up with what is now called Gatorade and tried it out on the Florida B team (the second-stringers) during an intrasquad scrimmage with the A team of the Gators. Needless to say, the B team, overmatched in talent but not suffering from dehydration or salt imbalance, beat the A team. The legend of the performance-enhancing abilities of this simple drink spread rapidly in the high schools and colleges of Florida and is now found on the bench at nearly every college and professional football game in the country.

## CHAPTER 5 Oxygen—The Breath of Life

**1.** Although not actually a physiologist, Joseph Priestley (1733–1804) made such a momentous discovery in chemistry that it revolutionized physiology as well—the "air" that he is talking about here is oxygen. This great Renaissance man didn't confine his thinking to chemistry and, like many early scientists, wore many hats, including those of philosopher, theologian, historian, and educator, to name a few. Born to humble parents in Yorkshire, England, Joseph was raised first by his grandparents and later by an aunt. He showed early signs of genius while mastering numerous languages, including Hebrew, Arabic, Greek, Latin, German, Syrian, and several Romance languages. As a minister and historian, he published several notable works.

In the course of his studies on the "goodness" or "badness" of air (that is, air with and without oxygen, respectively), Priestly recognized that plants can restore "goodness" to bad air, thus laying the groundwork for our present understanding of how plants produce oxygen. He also recognized that carbon dioxide could be dissolved in water and produce carbonation, as in natural mineral springs (and, today, in champagne and soda pop). Most important, however, he found that heating mercuric oxide resulted in the production of an invisible gas that could keep a mouse alive in an enclosed air-tight chamber. This invisible gas could also permit a flame to burn and appeared to make up about one-fifth of the gas in normal air (a correct assumption; the actual composition of air is 20.95 percent oxygen, the rest mostly nitrogen). This great discovery paved the way for all of modern respiratory physiology and medicine.

Priestley led a fascinating and, at times, horrific life. Eighteenth-century scientists rarely sequestered themselves in a so-called "ivory tower," and their views were often the subject of public controversy, partly because many of these academicians were socially and politically well connected. In Priestley's case, his ideas about religious doctrine began to change as he got older, and he openly questioned traditional interpretations of such basic tenets of Christianity as the Trinity. For such an intelligent man, it was a pretty foolish thing to do in those days. He was forced to flee England and watched his home, laboratory, and library go up in flames at the hands of a riotous mob. This was not as bad as the fate that befell his French friend and colleague Antoine Lavoisier, who might be regarded as a co-discover of oxygen and met a premature end via the blade of a guillotine during the French Revolution. Priestly emigrated to America and took up residence in Pennsylvania, where he began to repair his reputation and continue his studies. Long a friend of Benjamin Franklin, who he met in England, he also became a friend and adviser to two U.S. presidents, John Adams and Thomas Jefferson, and founded the first Unitarian church in the New World. In 1973 a newly discovered crater on Mars was named the Priestley crater in his honor.

**2.** Take a look at someone's neck when they're lying down, and you'll clearly see pulsations in the neck. It is often thought that these are the carotid arteries, but in fact that's not true. Arteries do indeed have a pulse, as blood pressure rises with the beat of the heart, and lowers as the heart relaxes, but most arteries are buried deep under the skin and not readily visible. If they were close to the skin, it would present the danger that a simple surface wound could rupture an artery, a situation that could have disastrous consequences. There is so much pressure in arterial vessels that when cut, an artery produces tremendous bursts of blood that can spurt several feet in the air. A ruptured artery can quickly lead to circulatory shock and death. But a vein has almost negligible blood pressure; normally, little long-term damage occurs following a cut vein. When you cut yourself, the blood you see is usually coming from veins, not arteries. Instead of spurting out under high pressure, it dribbles out of the low pressure veins, allowing time for clotting to take place. That, incidentally, is one reason why so few people are actually successful at committing suicide by slicing their wrists. They only cut through veins, and the blood dribbles out slowly enough that the body can cope with the loss.

The pulsing you see in a person's neck is actually from the two jugular veins, which lie very close to the skin in the neck. Like all veins, these do not actually have a true pulse. It looks like a pulse for several complex reasons having to do with the way in which blood flows back from the head into the heart. The jugular veins drain blood from the head and brain and dump it into the right side of the heart, causing the right atrium (or filling chamber) to bulge as it fills with blood. This sends a wave of turbulence back up the jugular vein that is seen as a small pulsing motion. Later, as blood leaves the filling chamber of the heart and enters the pumping chamber, or ventricle, two additional waves of turbulence are generated, resulting in two more pulse-like motions in the neck. These are best seen when someone is lying down, because when standing, gravity tends to dampen the upward wave.

The jugular veins are huge and accessible for surgery. Many scientists studying physiology need to obtain repeated blood samples from a laboratory animal, like a rat. One way to do this is to place a short, flexible piece of hollow tubing (a cannula) into the jugular vein of laboratory rodents, seal the incision site, and tie off the cannula. The veins are so large and near the surface that the entire procedure, done under anesthesia, takes only about five minutes in the hands of a skilled surgeon. Blood samples can then be withdrawn from an extension on the cannula through a port in the top of the rat's cage. In this way, blood samples can be taken while the animal sleeps, and the rat is never aware of the investigator's presence—a remarkably stress-free way to approach tricky physiological experiments.

Surprisingly, some people have had their jugular veins severed and lived to tell the tale. Although large, the jugulars have very little pressure in them, and people can survive a cut jugular with prompt medical attention. On the other hand, if the carotid artery is severed, it almost surely means immediate death.

**3.** We call the brain stem a "primitive" part of the brain because it arose very early on in vertebrate evolution. In fact, the brains of many lower vertebrates are not much more than a brain stem and a few additional bits. It's not until you get to reptiles that a true cortex begins to appear, and not until mammals do you find a true, well-developed frontal lobe, or neocortex.

The part of the brain stem that controls breathing is actually composed of several regions that collectively reside in the pons and medulla oblongata areas. If these areas are damaged—such as from a stroke—then the work of breathing would depend upon an artificial device called a respirator. We say that these areas of the brain control the vegetative functions of the body—respiration, digestion, glucose balance, and cardiovascular activity.

**4.** The nerves that activate the rib muscles are the intercostal nerves, and the large nerve that activates the diaphragm is the phrenic nerve. Rhythmic signals from the brain stem stimulate these two sets of nerves, and the resulting muscle contractions expand our chest. Whenever a gas-filled chamber is expanded, the pressure of gas in that chamber decreases (Boyle's Law; see Chapter 4, note 4). Thus, if the air pressure in the lungs decreases as the chest expands, it creates a downhill pressure for air to flow from outside the body and into the lungs. When the nerve signals turn off, the muscles relax and the lungs and chest wall snap back into place much like a stretched rubber band.

**5.** For a short history of the erythropoietin story in athletics, see I. Austen, "Blood Feud," *Cycle Sport* (August 1996).

**6.** A major job of the spleen is the formation of white blood cells, which are part of the body's immune system, but it also has a sieve-like structure that filters out dead or abnormal red blood cells and removes them from the circulation. Only in some animals does the spleen act as an important reservoir of good red blood cells, to be used in times of emergency. In adult human beings, the spleen is not even essential for life. People who have had their spleens removed—often because of traumatic injury—manage to function very well. The spleen is probably much more important in newborns, in whom the immune system is still developing.

**7.** Normally, small amounts of acid can be handled by the kidney, which has the means of returning pH to normal. Higher amounts of acid can overwhelm the kidney, at least temporarily. Most enzymes in the body are very sensitive to changes in pH. Thus, lowering the pH (making the blood more acidic), may alter the effectiveness of countless enzymes, and this is what creates such havoc with normal health in such a situation. By the way, making the blood more alkaline (higher pH) can be just as bad as making the blood more acidic. It works both ways, because enzymes only work properly within a very narrow pH range.

## CHAPTER 6 Life Under Pressure

**1.** Allow me to digress a moment on the subject of William Harvey, one of the most influential scientists of the last millennium and, incidentally, one of my all-time heroes. Harvey doesn't share the sort of instant name recognition of Copernicus or Galileo or even modern-day scientists like Watson and Crick of DNA fame, but he shaped our modern view of medicine as much as any individual.

Harvey was an English physician born in Folkstone, Kent, in 1578, and received his degrees from Cambridge University. Much of his greatest work took place at the

University of Padua in Italy (one of the great centers of the study of anatomy, physiology, and medicine for centuries). A contemporary of Shakespeare, Cervantes, and many other great molders of the Enlightenment, he was official physician to James I and his successor, Charles I. His treatise *Exercitatio anatomica de motu cordis et sanguinis in animalibus* (Essay on the motion of the heart and blood in animals) caused quite a stir. In it, he demonstrated that animals have a fixed quantity of blood that is reused over and over again through a closed, self-contained circulation. Through careful experimentation, he deduced that the human heart must pump out about 10 pounds of blood each minute (which is very accurate). This was far too much to have been converted from food, because it is obvious that no one eats that much food in a day, let alone in a minute. Thus, he finally put to rest the Galenist view of the circulation (see Chapter 1), in which the liver converts food into blood, and opened the door for modern investigation of the nature of the circulation. This was no small feat and required extensive experimentation and surgical skill to prove.

Harvey's other great contributions have been largely forgotten except by those who work in the field, but they, too, were very important in their own right. For example, he was among the first to put forth the view that all living creatures—including man—begin life as an egg. He deduced that germ cells were not infinitesimally small versions of fully grown animals, as previously thought, but were completely undifferentiated. Prior to Harvey, most physicians accepted the idea that germ cells were perfectly preformed creatures that expanded as they grew into fetuses. Working on developing chick embryos (still a common animal model for the study of development), Harvey made the intellectual leap that factors of some kind were involved in shaping an organism from an immature, unformed state into a complex, fully formed one. Not everyone accepted this argument or his ideas about the circulation. The dean of the illustrious Paris Faculty of Medicine, for example, publicly called Harvey's ideas "absurd" and "useless." Not surprisingly, the name of this particular dean is lost to all but historians.

Harvey's other contributions related to the origin of disease and embryology are innumerable. When he died in 1657, at least one great mystery eluded him, namely, what was the purpose of the circulation in the first place? His idea was that blood was circulated to transfer heat from the heart—where he believed it was generated—to the distant parts of the body. This was, in a sense, not entirely incorrect, because heat exchange and the circulation are intimately tied together, but he missed the significance of the blood in carrying oxygen and nutrients around the body. That idea would come later.

Incidentally, Harvey was unable to actually visualize the connections between arteries and veins, called capillaries, although he was the first to deduce their existence. Marcello Malpighi (1628–1694), an Italian anatomist born the same year that Harvey published his famous treatise, confirmed the existence of capillaries in 1661 using the newly invented microscope. And to give proper credit, Harvey was not actually the first person to suggest that the Galenist view of the heart was wrong and that a circulation of sorts exists. Miguel Serveto (later, Michael Servetus, 1511–1553) of Spain had suggested this idea (but not proved it) much earlier. Unfortunately for him, the world wasn't quite ready to receive such knowledge, and he was imprisoned by the Spanish Inquisition. Of course, he didn't help his cause by publishing a treatise that questioned the validity of the Trinity and other doctrines of Christianity. After escaping from jail, Serveto was hunted down and burned at the stake by order of none other than John Calvin. Fortunately, the perils of studying physiology are not what they once were.

As the quotes at the beginning of this chapter illuminate, Harvey was a diplomatic man and knew what side his bread was buttered on. The last line from one of his historic treatises, addressed to the president of the Royal College of Physicians, was, "Think kindly of your anatomist."

For more on William Harvey, see Walter Pagel, *New Light on William Harvey* (Karger Press, 1975). A good English translation of Harvey's most famous treatise is by C. D. Leake (Charles C. Thomas, 1949); this translation is accompanied by numerous historical footnotes and interesting biographical information.

**2.** The rhythmic beating of both the jellyfish bell and our heart muscles is under the control of the nervous system. There are pacemaker nerves in the jellyfish bell that fire off at regular intervals, and speed up when the jellyfish needs to increase its activity. In fact, the electrical characteristics of the nerve signals that are generated by the jellyfish pacemaker look very much like those produced in our heart. Furthermore, in both cases the entire group of muscle cells contracts at the same time. The nerve impulses from the jellyfish pacemaker spread their signals over the whole bell almost instantaneously, so that the entire bell contracts in a coordinated way. The same thing happens in our heart; the pacemakers generate an electrical signal that immediately sweeps over the entire heart. This ensures that the millions of heart muscle cells behave as if they are all one giant muscle cell, and contract together.

**3.** The athletes with the largest hearts were those in water polo, rowing, and cycling sports. Of those tested, the least aerobically challenging sports were diving, Tae kwon do, and team handball. Not surprisingly, equestrianism was low on the list, too, but was higher than alpine skiing! Look for the complete reference in the *New England Journal of Medicine* 324 (1991): 295–301.

**4.** Most of us experience a mild form of orthostatic hypotension on rare occasions. It usually is only fleeting upon standing; especially, as I've said, when we're dehydrated on a hot day and have been sitting or lying down. As we suddenly stand up, blood falls due to gravity and the pressure of blood reaching the brain decreases. This causes the sensation of light-headedness and, if we're not careful, we can faint. We normally don't, however, because the baroreceptors in our necks sense the loss of pressure and initiate a cascade of neural and hormonal responses that very rapidly bring pressure back up. One of these hormones, renin, is made in the kidney and released into the bloodstream whenever the pressure receptors sense that pressure is lower than normal. Once in the blood, renin acts on a molecule called angiotensinogen, which is made by the liver. The angiotensinogen is thereby converted into a smaller molecule called angiotensin I, which in turn is further broken down into the final, active molecule called angiotensin II, which does a variety of things. It makes animals (and presumably, us) thirsty so that they drink more fluids, which restores pressure; it stimulates blood vessels to constrict, which raises the pressure within them; and it activates production of yet another hormone, called aldosterone. Aldosterone is made by the adrenal glands and is essential for life. It acts on the kidney to help the kidney tubules reabsorb water that would otherwise be lost in the urine. Thus, angiotensin II is a vital hormone that triggers numerous mechanisms, all of which are designed to increase circulating blood volume and pressure. If you had no ability to make angiotensin II, you would be much more likely to suffer from serious orthostatic hypotension.

Besides the hormonal responses that occur when we stand and get light-headed, the branch of the nervous system known as the sympathetics kicks in immediately when pressure reaching the brain falls even slightly. The nerve endings of the sympathetics release norepinephrine onto blood vessels, which makes them constrict, thus elevating pressure. This response is so rapid compared to the hormonal response that it can be considered to be almost instantaneous.

## CHAPTER 7 Bat Wings and Elephant Ears: Keeping Cool

**1.** The formula to convert degrees from the Fahrenheit scale to the Celsius is to subtract 32 and multiply the result by 0.55. Thus, for 100 degrees F, we subtract 32 from 100, or 68, multiplied by 0.55, which equals 37.4 degrees C. The story of how the Celsius scale came about is surprisingly interesting. Anders Celsius was an eighteenth-century Swedish astronomer who, along with others, was interested in establishing an accurate way of recording temperature. He decided to use the melting and boiling points of water as two standard temperatures and divided the intervening temperatures into 100 equal units. What he actually measured was the height of a column of mercury in a glass tube that was heated to varying degrees, much like today's thermometers. He also correctly noted that at different elevations, water boiled at different temperatures, and reasoned from this that his thermometer could take the place of barometers. Strangely, however, in calibrating his thermometer he set the boiling point of water at 0 degrees, and the freezing point at 100 degrees. No one is sure why he set it up in this odd way, nor who had the common sense to switch it around in later years. Most of the world—and all of science—follows the Celsius scale because it's less cumbersome than the Fahrenheit scale, in which water boils at the awkward temperature of 212 degrees and freezes at 32 degrees.

**2.** When I refer to fats, I'm specifically talking about the so-called fatty acids. These are long chains of carbon molecules bonded one to the other. If one or more of the bonds is modified into a double bond, we say the fatty acid is now unsaturated. As the number of these double bonds increases, the fatty acid becomes more unsaturated, and it remains in an oily state at lower and lower temperatures.

**3.** Incidentally, the change in fatty acids to an unsaturated state also contributes to the ability of a fish to cope with the hydrostatic pressure of water. Unsaturated fatty acids are not as easily squeezed by the pressure exerted outside an animal's body underwater. They tend to retain their shape better, and therefore preserve the integrity of the membranes around all of the body cells. Because these membranes are made up of roughly 40 percent fatty acids, this is very significant.

## CHAPTER 8 Sensing the World Around Us

**1.** Among the many serious effects of alcohol on the body is its depressant effect on the central nervous system, especially a part of the brain known as the cerebellum (literally, "little cerebrum"). Within the cerebellum is the balance center of the brain. Disruption of cerebellar function due to a tumor or disease creates dizziness and other symptoms. Most common is the loss of fine motor control. Picking up a pencil or reaching for any small object becomes a trial. In addition, standing on one leg (especially with eyes closed), walking a straight line, or touching the tip of the nose with the fingertips become equally difficult. That's because the signals from the proprioceptors in the arms, fingers, legs and toes send their signals to the cerebellum. Thus, the symptoms of inebriation and cerebellar disease are in many ways quite similar. In fact, the same tasks a policeman puts a suspected drunk driver through are those that form part of any general neurological exam in a doctor's office. Not all balance problems are indicative of cerebellar disease, however. Millions of people suffer from an awful disease known colloquially as vertigo and clinically as benign positional vestibulopathy. In these individuals, unidentified defects in their inner ears send inappropriate messages to the cerebellum, making them feel dizzy even when they are motionless. In this case, it is the inner ear and not the brain that is malfunctioning.

**2.** Taste buds are not only located in the tongue but are also found in the pharynx (throat) and esophagus. These receptors may have evolved to ensure that food would not just be chewed and spat out but swallowed (think how much more satisfying it is to swallow ice cream rather than just lick it and spit it out).

## CHAPTER 9 Stone Age Stress and Coping with Change

**1.** Selye's seminal work was published in the prestigious scientific journal *Nature* in 1936, when he was just 29 years old. He described in exquisite detail the histological changes in the appearance of virtually all body tissues and organs after exposing an animal (a rabbit, in this case) to one or another type of stress. The number of changes he observed are too numerous to list here, but of special significance were the observations that all stresses, or "noxious agents," as he first called them, regardless of their origin, resulted in the same internal changes. Notably, the adrenal glands (source of cortisol) enlarged, the reproductive glands got smaller, and the lymph tissue shrank. By today's standards, his publication appears at first glance to be rather shallow, filling up less than a single page of the journal, with no charts, tables, or statistical treatments. But it is one of the most celebrated publications in modern physiological research, because of its accuracy and attention to detail, and because it was responsible for launching stress research as a valid scientific discipline. Selye, who died in 1982, went on to publish many important papers in the field but was never awarded the Nobel Prize for his efforts. Nonetheless, later investigators did reap the benefits of his early work, and to date 21 scientists have been awarded or shared Nobel Prizes for work related to hormones, many of them directly or indirectly related to stress research. One of them was the American Philip Showalter Hench (1896–1965) of the Mayo Clinic. In 1948 he showed that glucocorticoids had anti-inflammatory actions and that this could be used to clinical advantage to treat arthritis, graft rejection, and other syndromes. He treated 14 severely disabled arthritic patients with cortisone (another glucocorticoid) and demonstrated rapid and dramatic recovery. In 1950 he shared the Nobel Prize with two other scientists whose earlier work led to the discovery of the adrenal steroid hormones.

**2.** We're used to thinking of cholesterol as the great evil of blood-borne compounds. In large amounts, it certainly can have serious consequences, such as contributing to the formation of atherosclerotic plaques in arteries, but it is also true that without cholesterol we'd be in big trouble. It is an important component of the membranes that surround all living cells and, in the present context, is the precursor molecule from which all steroid hormones are derived. Thus, estrogen, progesterone, testosterone, vitamin D, cortisol, and aldosterone (involved in salt and water balance) are all directly made from cholesterol. The backbone structure of cholesterol is retained in all these hormones, and thus they are chemically quite similar in nature but very different in function.

**3.** The importance of vasopressin to normal health is underscored by the disease known as diabetes insipidus, or DI, as it's usually called by scientists and clinicians. The only similarity between DI and the other kind of diabetes (mellitus), is that large amounts of urine are formed by the kidney when the disease is not controlled.

In the absence of vasopressin, water cannot be properly retained by the kidney as it filters blood and removes the toxic waste products of metabolism. The water is excreted in huge amounts, and it becomes nearly impossible to drink enough liquids to compensate for what's lost. A person with DI develops severe dehydration and a life-threatening imbalance in the salt level in the blood, because the water content drops so drastically. Seizures follow and, if not corrected, the disease can be fatal. Modern research has led to the development of a long-acting vasopressin molecule (known as DD-AVP) that can be inhaled in much the same way as over-the-counter decongestant sprays. The blood vessels of the nose are unusually leaky and can absorb the vasopressin right into the blood system, thus allowing the person with DI to lead a normal life. Vasopressin sprays are also occasionally used at nighttime to help prevent bed-wetting in older children and are remarkably effective.

**4.** Harvey Cushing, after whom the disease is named, lived from 1869 to 1939 and did most of his greatest work at Harvard University. He was one of the founders of modern-day neurosurgery and was the first to successfully perform surgery on diseased pituitary glands. Cushing was also an author, and, in addition to his scientific and medical accomplishments, which are legendary, he received a Pulitzer Prize in 1925 for his biography of the British clinician Sir William Osler, a widely regarded medical educator. The disease that bears his name is usually caused by a tumor within the pituitary gland that makes excess amounts of ACTH, which in turn causes the adrenal glands to become overstimulated. The high amount of cortisol that results causes hypertension, muscle and bone wasting, thinning of the skin, immunosuppression, weakness, and a very characteristic redistribution of fat in the body. Fat accumulates in the face and the back of the neck, giving sufferers of the disease a characteristic moon face appearance. The treatment for the disease is to surgically remove the tumor if possible.

**5.** For more on Cushing's disease and Addison's disease, contact the National Adrenal Diseases Foundation in Great Neck, NY.

**6.** Like anything else, exercise is good for you in moderation. Among the many benefits of exercise is the development of higher levels of so-called "good" cholesterol, which seems to provide resistance to coronary artery disease. But chronic, strenuous exercise can be debilitating. A dramatic example of how stressful exercise can be is seen when laboratory animals are trained to strenuously exercise for long periods of time (for example, a rat on a running wheel). Eventually, the reproductive system of the rat shuts down, sex hormones decline in the blood, the ovaries or testes shrink, and the rat becomes infertile. It's as if the body has decided that its metabolism is so seriously overwhelmed that any unnecessary function must be shut down until the animal gets its systems back to normal. Reproduction may be essential for continuing the species, but it is energetically costly and isn't needed for survival of the individual.

Does this happen in people? It does. For example, the amount of training required by elite young gymnasts or ballet dancers is so intense that it constitutes a serious stress to the body. There is little energy left over for the metabolically demanding functions of reproduction. Thus, just as the rat stopped having its own version of a reproductive cycle, adult ballerinas stop having menstrual cycles, a condition known as exercise-induced amenorrhea. During this time, ballerinas and any other similarly active women are infertile. The situation for the gymnast, on the other hand, is worse, because the stress usually starts earlier in life. As a consequence, they fail to enter puberty at the normal time, and once they begin puberty, it develops slowly. The first menstrual period in young elite gymnasts is usually very late, the growth spurt is delayed, and the voice remains juvenile; the result is small, 16-year-old, squeaky-voiced, prepubertal girls. As a physiologist, I cringe every time I see these young athletes performing on the world stage, as in the Olympics. It is a grossly abnormal situation, and one can only imagine what psychological effects it must have as well.

## FURTHER READING

There are many good books and reference articles on the subject of stress, feedback, glucocorticoids, and the adrenal gland. For a clear and thorough description of this process and stress in general, see Robert Sapolsky, *Why Zebras Don't Get Ulcers: A Guide to Stress, Stress-Related Diseases, and Coping* (W. H. Freeman and Company, 1994). For technical information that covers most of the relevant bases, including Addison's and Cushing's diseases, I suggest C. R. Kannan, *The Adrenal Gland* (Plenum Press, 1988). Another good book on stress, particularly the psychological aspects of stress, is M. H. Appley and R. Trumbull, eds., *Dynamics of Stress: Physiological, Psychological, and Social Perspectives* (Plenum Press, 1986). Perhaps the most widely cited technical article on feedback and stress is M. Keller-Wood and M. F. Dallman, "Corticosteroid inhibition of ACTH secretion," *Endocrine Reviews* 5 (1984): 1–24. Although a bit outdated, it is nonetheless a valuable survey of negative feedback in the stress response.

# INDEX

Page numbers in *italics* indicate illustrations;
*n* indicates a note.


abortion, 156*n*, 157*n*
acid taste buds, 121
acidic blood, 55, 175*n*
action potentials, 44, 123–124
Adams, John, 172*n*
adaptation
  to change, 5
  taste buds, 121
Addison, Thomas, 136–138
Addison's disease, 33, 136–138, 185*n*
adenosine triphosphate (ATP), 50, 94, 101, 102, 132, 161*n*, 165*n*
adipocytes, 29
adrenal glands, 23, 127, *135*
  and Addison's disease, 136–138
  and aldosterone, 179*n*
  autoimmune destruction of, 136
  and cortisol, 129, 135
  Cushing's disease, 184*n*
  and epinephrine, 128
  of seal, 60
  and steroid hormones, 165*n*
  and stress, 126, 182*n*
adrenaline, 128
  *see also* epinephrine
adrenocorticotropic hormone (ACTH), 129, 135, *137*, 184*n*
adult-onset diabetes, *see* non-insulin-dependent diabetes mellitus (NIDDM)
air
  composition of, 171*n*
  nitrogen content, 65
  oxygen content, 53, 65
  pressurized, 61
air pressure
  levels of, 65
  in lungs, 174*n*
air sacs (alveoli), *67*, 90
  birds, 69–70
alcohol
  and dehydration, 44
  effects on body, 181*n*
aldosterone, 179*n*, 183*n*
alimentary canal
  of fish, 49
alkaline blood, 175*n*
allergies
  and heart rate, 83
  and loss of smell, 114
  and shock, 88–89
altitudes, high, vii–viii, 56, 64–68
alveoli, *see* air sacs
Alzheimer's disease, 24
amenorrhea, 185*n*
AMGEN, Inc., 36
amino acids, 32
amphibians
  body temperature, 92
  lungfish, 46, 167–168*n*
  toleration of freezing, 92
anaerobic respiration, 59–60
anatomy, vii
  scaling, 161*n*
Andes
  barometric pressure, 65
  survival in, viii, 4, 64
anemia, 22, 56, 138
angiotensin I and II, 85, 179*n*
angiotensinogen, 179*n*
animal experimentation, 152–153, 155

# INDEX


animals
  changes in form and evolution, 3
  and feel, 121–124
  growth of and mass and surface area, 10–11
  interrelatedness of, vii
  and seeing, 115–120
  sensory perception of, 115–116
  species classification, 156*n*
  size of and eating, 6–17
  survival challenges, 4
  and water stores, 4
  *see also* fish; mammals; *and individual animals, e.g.,* bats, mice, *and* whales
animals, cold-blooded, 92
  body heat, 17
  body temperature, 94–95
  breathing, 55
  and cold climates, 102
  *see also* fish; reptiles
animals, desert
  metabolic water, 167*n*
  and water, 45
animals, domesticated
  obesity in, 20
animals, large
  body heat loss, 13
  heat retention, 97
  life span, 17
animals, nocturnal
  smell and sound, 106
animals, small
  body temperature, 11–12
  breathing rate, 14, 55
  circulatory system, 16
  and eating, 6
  heart rate, 81, 82
  and heat loss, 11, 162*n*
  life span, 17
  metabolic rate, 10, 16–17
  *see also* bats; birds; mouse; shrew
animals, warm-blooded, 2, 92, 96–102
  breathing, 55
  eating, 96–97
  heat retention, *100, 101*
  and hibernation, 103
  overheating, 97
  *see also* mammals
animals, warm-bodied, 96
*Anolis* lizards, 147
anorexia, 30
anosmic people, 113–114
Antarctic icefish, 92–93
*Antechinus*, 131–132
antelope, 128, 130
antidiuretics, 130
antifreeze molecules, 92–93
antihistamines, 89
anxiety, 126
  primitive man, 139
aorta, 81, 87, 89
appetite control, 27
arachnids
  lungs, 168*n*
arid regions
  and dehydration, 139
  and sufficient water, 45
Aristotle, 156*n*
arteries, 14, 77–78, *80*, 81, 158*n*
  chemoreceptors in, 54, 55
  and circulation, 75
  location of, 172*n*
  pressure in, 87, 172*n*
  tuna, 95–96
  weakening of, 86
arterioles, 78
arthritis, 135, *137*, 139, 183*n*
Asclepius, 156*n*, 157*n*
asthma, 89, 153, 155
atherosclerosis, 19, 20, 38, 86

# Index


athletes
  blood doping, 56–57
  calorie consumption, 7
  cardiovascular fitness, 77
  heart size, 178*n*
  *see also* sports
ATP, *see* adenosine triphosphate
atria, 75, 79, 173*n*
Atrovent, 155
auditory cortex, 113, 124
aurochs, 140
autoimmune disease, 136
  diabetes, 33–34

bacteria
  and white blood cells, 106
balance, 106, 181–182*n*
ballet dancers, 185*n*
bar-headed goose, 69–70
barometric pressure, 65
baroreceptors, 86–87, 179*n*
basal metabolic rate, 7–9
basking, 94
bat, brown (*Myotis lucifugus*), 39
bat, bumblebee, 12
bat wings
  blood flow in and heat dissipation, *100*
  and body heat, 97–98
  surface area of, 91–92, 98
bats
  body temperature, 12
  brown fat, 99
  ears, 145
  echolocation, 112–113
  energy expenditure, 20
  fat stores, 39
  food consumption, 11, 12
  heart rates, 15
  hibernation, 19–20, 39, 103
  and oxygen, 4
  torpor, 103
Beard, Mary Ritter, 6
bears, 103
bed-wetting, 184*n*
bee stings
  and shock, 88, 89
bends, the, 2, 61–64
benign positional vestibulopathy, *see* vertigo
beriberi, 22
Bernard, Claude, 160*n*
bicarbonate, 55
biology, molecular, vii, 150
birds
  air sacs, 69–70
  bones, 69
  breathing, 69–70
  cranial cavity, 118, *119*
  and eating, 6
  energy requirements for flying, 19
  eyes, 114, 116, 118–120
  fat stores and migration, 38–39
  heat dissipation, *100*
  keeping warm, 99
  lungs, 69–70
  orientation and earth's magnetic field, 123
  overheating, 97
  photoreceptors, 116–120
  physiological evolution, 142
  respiratory system, 69–70
  retinas, 118, 120
  sense of smell, 114
  sight, 106, 114, *119*
  stress responses, 127
  temporal resolution of, 118
  trachea, 70
birds of prey
  eye physiology and sight, 117–120

bitter tastes, 121, *122*
bladder, 106
blindness, 38
blood
  acidic, 55, 175*n*
  alkaline, 175*n*
  antifreeze molecules, 92–93
  and body warmth, 99
  carbon dioxide in, 54–55
  cardiac output, 15–16
  chemoreceptors, 106
  dissolved gas in, 61–62
  early ideas about, 158–159*n*
  hematocrit, 169*n*
  and lactic acid, 59
  low oxygen pressure, 68
  pH balance, 59, 175*n*
  salt levels, 4–5, 41–42, 47, 48, 51
  in seals, 63
  warming of in tuna, 95–96
  water content, 42, 48
  and water loss, 48, 169*n*
  *see also* circulatory systems; hemoglobin
blood circulation, 5, 63–64, 71-74
  discovery of, 176*n*, 177*n*
  fish, 46
  gravity and pressure, 3, *76*, 78, 83–84, 179*n*
  and heat, 98, 99, *101*
  and oxygen transport, 4
  and shock, 88–89
  vessels, 77–81
blood content
  in diving mammals, 56, 57
  of mountain dwellers, 65–66
blood doping, 56–57
  natural, 58
blood oxygen, low, *see* anemia
blood pressure, *80*–84, 158*n*
  in arteries and veins, 172*n*
  baroreceptors, 86–87
  and blood loss, 87
  and blood volume, 83
  and body weight increase, 19
  and decongestants, 83
  and defense mechanism responses, 127
  different levels of in the body, 83
  drop in, 179–180*n*
  and excess fluid, 52
  giraffe, 2–4, 75
  and gravity, *76*, 83–84
  jugulars, 173*n*
  and light-headedness, 179*n*
  in lungs and mountain climbing, 66–67
  maintaining constant levels of, 4
  measuring, 84–85
  and stretch receptors, 106
  veins, 79
  and water loss, 47–48
blood pressure, high, *see* hypertension
blood pressure, low, *see* shock
blood sugar, 136
blood vessels, 74, 75
  and defense mechanism responses, 127
  and flushed appearance, *101*
  and heat dissipation, *100*
  and heat radiators, 98
  and keeping warm, 99
  nose, 184*n*
  tuna, 95
  water content, 42
blood volume
  and angiotensin II, 179*n*

and blood pressure, 83
hypovolemia, 169*n*
bloodstream, *see* blood circulation; circulatory systems
blue whale, 16
blushing, 98
BMI, *see* body mass index
body heat
cold-blooded animals, 17
radiators, 97–98
body mass
and breathing rate, 16
and heart rate, 16
and metabolic rate, 16, 161–163*n*
and surface area, 10–11, 97, 162–163*n*
body mass index (BMI), 36–38
body size
and eating habits, 6–17
and heart rate, 15–16
and heart size, 15, 75
and heat loss connection, 16–17
and lung size, 14, 56, 163*n*
and metabolic rate, 9–11, 17
mountain dwellers, 65
scaling, 161*n*
body surface area, *see* surface area
body temperature
and bat wings, 92
cold-bloodedness, 92
and elephant ears, 92
and fever, 31
fish and water temperature, 95, 103
fluctuations in, 160*n*
and heat radiators, 92
and hibernation, 103–104
maintaining, 11
and overheating, 97
and temperature detectors, 98
warm-bloodedness, 92
body weight
and calorie consumption, 7
and capillaries, 47
and heart size, 161*n*
regulation of, 28–31
and water, 42
bone loss, 135
bone marrow, 56, 88
bone wasting, 184*n*
bones
birds, 69
book lung, 168*n*
Boyle's Law, 168*n*, 174*n*
brain
and action potentials, 123–124
auditory center, 113
evolution of, 143, 146
balance center, 181*n*
and baroreceptors, 87
and blood pressure, 179*n*
corticotropin-releasing hormone, 165*n*
eating behavior mechanism, 22–27
as excitable tissue, 44
lesioning experiments, 23–25
limbic system, 128
and negative feedback, 133
neurons, 25–26
neurotransmitters, 164–165*n*
nuclei, 22, 23, 25, 164*n*
and oxygen, 53, 68
and sensory stimuli, 111, 124
signaling process of, 44
stimulation experiments, 25
and stress, 128, 129

temperature detectors, 98
see also cerebellum; forebrain; hypothalamus; neocortex

brain cells
death of, 42, *43*, 53
at high body temperature, 97
nuclei, 164*n*
and oxygen, 53, 54
and salt level in blood, 41
and sodium concentration, 52

brain damage, 24–25, 97, 124

brain edema, 68

brain regions
identifying functions of, 24–25

brain stem, 128, 174*n*
feedback, 133
and oxygen detection, 54
and stress, 129

breath, shortness of, 90

breathing, 53–55
birds, 69–70
brain stem control, 174*n*
and defense mechanism responses, 127
gills, 50
humans, 70
marine mammals, 55–60
*see also* hyperventilating

breathing, water, 47, 48

breathing rate, 14, 55
and body mass, 16
tuna, 95

bronchioles, 89, 153

bronchoconstriction, 89

brown bat (*Myotis lucifugus*), 39

brown fat, 99–100

bumblebee bat, 12

Cade, Robert, 169–170*n*

caffeine, 83

calcium, 44

calories
average consumption by humans, 7
burning of and oxygen consumption, 8

cancer
and Addison's disease, 138

Cannon, Walter, 127, 160*n*

capillaries, *78–80*, 177*n*
in fish, 46–47
and heat radiators, 98
in lamellae, 168*n*
in lungs, *67*, 68
in mountain dwellers, 66

carbon dioxide, 171*n*
in blood, 54–55
and breathing, 54
circulatory system, 79, *80*
pressurized, 61
as waste product, 163–164*n*

carbon molecules
fatty acids, 180–181*n*

carbonation, 61, 63, 171*n*

cardiovascular activity, 174*n*

cardiovascular diseases, 84–85
*see also* hypertension

cardiovascular fitness, 77

cardiovascular medicine, *72*, 73

cardiovascular system
mammals, 14–15
mountain dwellers, 66

carnivores, 22

carotid arteries, 54, 87, 172*n*, 173*n*

cat
eyes, 144

catabolism, 129

CCK, *see* cholecystokinin

cell death, 42, *43*, 53

cells
- chemoreceptors, 54, 55
- freezing of, 93–94
- and fuel-burning enzymes, 11
- intracellular and interstitial fluid, 42
- oxygen-detector, 54

Celsius, Anders, 180*n*
Celsius scale, 180*n*
central nervous system
- effects of alcohol, 181*n*

cephalopods
- circulatory systems, 74

cerebellum, 181*n*
Charles I, 176*n*
chemical reactions
- temperature and, 94, 95
- *see also* metabolic rate

chemoreceptors
- in blood, 106
- and carbon dioxide, 54
- and oxygen stores, 54, 55
- smell and taste, 106

chick embryos, 176*n*
children
- blood pressure, 84
- fat cells, 29
- heart rates, 81

chimps
- neocortex, 146

Chinese physicians, 159*n*
Chinese Restaurant Syndrome, *122*
cholecystokinin (CCK), 30, 32
cholesterol, 19–20, 129, 183*n*
- and atherosclerosis, 86
- and exercise, 185*n*

chronic fatigue syndrome (CFS), 138
circulation, *see* blood circulation
circulatory collapse, *see* shock, circulatory
circulatory system, *71–74, 80*
- pumps and vessels, 74–81
- seal, 63

citrus foods, 121
cold
- defined, 107

cold-blooded animals, *see* animals, cold-blooded
cold climates
- mammals in, 103–104

cold medications
- and blood pressure, 83

Coleman, Douglas, 28, 35
Coleridge, Samuel Taylor, 41
color receptors, 111–112
- bird retinas, 120

communication
- within and between species, 5
- within hypothalamus, 25

consciousness
- and shock, 87

corticosterone, 129
corticotropin releasing-hormone (CRH), 30, 128–129, 133–*135*, 165*n*
cortisol (hydrocortisone), 88, 126, 129–132, 134–135, 165*n*, 182*n*
- and Cushing's disease, *137*, 184*n*
- and extreme exercise, 141
- immunosuppressive effects of, 136
- and negative feedback, 132

cortisone, 183*n*
countercurrent heat exchanger, 95
cows, 6

cranial cavity
and eye size, 118, *119*
CRH, *see* corticotropin-releasing hormone
crocodile, 51, *93*
Cushing, Harvey, 184*n*
Cushing's disease, 135, *137*, 184–185*n*
cycling, 57, 178*n*

danger, *see* stress
DD-AVP molecule, 184*n*
deaths
from cardiovascular diseases, 85
decompression sickness, *see* bends, the
decongestants, 83
deer, 120
defense mechanisms
fight or flight response, 127–128
internal, and survival, 126–127
mammals, 127
dehydration, 44
and alcohol, 44
and blood fluid, 81
and cell death, 42
diabetes, 165*n*, 166*n*, 184*n*
from exercise, 83, 169–170*n*
primitive man, 139
and retaining water, 130
and salt imbalance, 52
stress of, 128
Denver
barometric pressure, 65
depth perception
and forward-facing eyes, *119*, 120
and lateral-facing eyes, 120
and two eyes, 113
diabetes insipidus (DI), 183–184*n*
diabetes mellitus, 33–35, 159*n*, 165–167*n*, 183*n*
causes, 19, 34
insulin injection, 32
Type I, *see* insulin-dependent diabetes mellitus (IDDM)
Type II (adult-onset), *see* non-insulin-dependent diabetes mellitus (NIDDM)
diaphragm, 54, 174*n*
diastolic pressure, 84
dichloromethane, 111
diet, 158*n*, 159*n*
and diabetes, 35
and hypertension, 85
and salt, 52
and taste buds, 121
dieting
metabolism slow down, 26–27
yo-yo effect, 26, 30
digestion, 174*n*
small animals, 13
dimetrodon, 104
dinosaurs, 90
blood pressure, 75
heat radiators, 104
directionality
of smell, 113
and stimuli receptors, 113
disease
and homeostasis, 5
from lack of eating, 22
and obesity, 38
dissection
and early physicians, 156*n*, 158*n*
diuretics, 68
diving
and bends, 61–63
and blood cell count, 58
diving mammals, *see* mammals, diving
dizziness, 165*n*

dog
eyes, 144
spleen, 58
taste, 120, *122*
dolphins
blood flow in, 64
and diving underwater, 56, 63, 64
neocortex, 146
Doyle, Sir Arthur Conan, 91
drinking, 5
drought
and retaining water, 130
drunk driver test, 109

ear, inner, 106, 182*n*
ear, middle, 106, 110
eardrum
and sound waves, 110
ears, 113
bats, 145
elephant, 2
mechanoreceptors, 110
owls, 145
sound arrival at, 113
eating, 5
drive to, 21–22
regulation of by Neuropeptide Y, 27
as source of water, 22
warm-blooded animals, 96–97
eating behavior mechanism, 22–27
eating habits
and animal size, 6–17
echocardiogram, 76–77
echolocation, 112–113
edema, *67*, 68, 78
electrical impulses, *see* action potentials
electrical sensing, 107
electrical signals
and sensory stimuli, 123
and sharks, 2, 4
electricity, animal, 44
electrolytes, 44
electromagnetic radiation, 107
electromagnetism, 1
elephant
body mass, 163*n*
body size and metabolic rate, 9–10
communication, 114
energy burning, 17
hearing, 145
heart size, 15
surface area of, 97
elephant ears
and heat dissipation, *100*
surface area of, 2, 91–92, 98
elevations, high, vii–viii, 56, 64–68
Emerson, Ralph Waldo, 1, 18, 40
emotions
and heart rate, 82
and stress, 128
emphysema, 55
endogenous morphine, 130
endorphin, 130, *131*
and extreme exercise, 141
energy
and adenosine triphosphate, 160–161*n*
and feedback, 132
of gill breathing, 50
and heat production, 7–8
and physiology, 2
energy consumption
tuna, 95
energy loss
bats, 20
fish, 47
energy production
and cortisol, 129
energy requirements, 6–7, 59
energy stores
warm-blooded animals, 96–97

environment
and evolutionary adaptations, 3
and oxygen transport, 4
and physical laws of nature, 5
sensing changes in, vii
environmental triggers
for diabetes, 34
enzymes
fuel-burning, 11
at high body temperature, 97
and metabolism, 160–161*n*
in mitochondria, 66
and sodium-potassium molecular pump, 101–102
epinephrine, 88, 89, 99, 126, 128
and brown fat cells, 100
fight or flight response, 128
in seals, 58
and stress response, *131*
Epo, *see* erythropoietin
equestrianism, 178*n*
erythropoietin, 56–57, 66, 174*n*
esophagus
fish, 46
estrogen
and cholesterol, 183*n*
and feedback, 132–133
oral contraceptive, 133
evolution
and changes in animal's form, 3
evolutionary adaptations
metabolism, 40
excitable cells, 52
exercise
and cholesterol, 185*n*
chronic and strenuous, 140–141, 185*n*
and dehydration, 83, 169–170*n*
and diabetes, 35, 167*n*
and flushed appearance, 98
and heart rate, 74, 82
and heart size, 75, 77
and hypertension, 85
and overheating, 97
and oxygen, 59
*Exercitatio anatomica de motu cordis et sanguinis in animalibus* (Harvey), 71, 176*n*, 178*n*
experimentation
animal, 152–153, 155, 157*n*
study of physiology, 149–155
eyes, *119*, 120
birds, 114, 116, 118–120
color receptors, 111–112
and depth perception, 113, *119*, 120
mechanoreceptors, 109

Fahrenheit scale, 180*n*
fainting, 82–84, 179*n*
fat cells, *see* adipocytes
fat stores
for migration or hibernation, 18–20, 38–39
fats, saturated, 102–103
fats, unsaturated, 102, 103, 181*n*
fatty acids, 180–181*n*
as energy carriers during stress, 129
and metabolism, 160*n*
unsaturated state, 181*n*
fatty mutant rat, 26–27
feedback, 132–135
feeding
control of, 30–31
and hypothalamus, 23
feeding centers (brain), 25

feel, 121–124
fertility
  exercise and suppression of, 141
fever, 31, 97, 156*n*
fight or flight response, 127–129, *135*
fish
  alimentary canal, 49
  antifreeze molecules, 92–93
  blood salt level, 47, 48, 50
  body temperature, 92, 95
  breathing, 169*n*
  coughing, 168*n*
  drinking seawater, 48–49
  and hydrostatic pressure, 181*n*
  internal water and salt level, 50
  kidneys, 50–51
  lamellae, 168–169*n*
  lateral lines, 106–107, 113
  and oxygen, 4, 45–48, 70
  prevention of freezing, 102
  and salt level of water, 2
  saturated fats, 103
  and stress, 127
  suffocation, 169*n*
  vibration detection, 106
  and water content, 4, 48–50
  *see also* gills; lungfish
flamingo, 98
flatworm, 107–108
flowers
  appearance of to insects and birds, *116–117*
fluids
  restoring during shock, 87, 88
flushed appearance, *101*
foie gras, 20
food
  ability to taste, 121
  intake, 13–14
  *see also* eating
forebrain, 143, 146
Franklin, Benjamin, 172*n*
freezing
  and antifreeze molecules, 92–94
  preventing, 102
Friedman, Jeffrey, 36
frog heart, 81–82
frontal lobe, *see* neocortex
fuel, *see* food
fuel burning
  body size and metabolic rate, 9–10
  enzymes, 11
  and heat replacement, 11
funding
  of physiology experiments, 151–152

Galen, Galenists, 71–72, 157–159*n*
gamma aminobutyric acid (GABA), 164–165*n*
gangrene, 38
gas
  dissolved, in blood, 61–62
gastrointestinal tract, 127
Gatorade, 169–170*n*
gazelle, 114–115
geese
  bar-headed, 69–70
  fat accumulation for migration, 18–19
genes
  for leptin, 36
  obesity, 21
germ cells, 176*n*
gill breathing, 50
gills, 46–47, 168*n*
  lamellae, 168–169*n*
  oxygen capture and delivery, 46–47

and salt content of blood, 50
surface area of, 46
giraffe
heart and blood pressure, 2–4, 75, *76*
neck, 3–4, 75, *76*
glucocorticoids, 129, 135, 136, 138, 183*n*
glucose, 165–166*n*, 174*n*
burning of in brown fat cells, 100–101
and chemoreceptors, 106
conversion to ATP, 102
and energy, 132
and metabolism, 160*n*, 161*n*
production of, 129
glycerol, 93
and metabolism, 160*n*, 161*n*
gods
and man's psychological stress, 140
goose liver, 20
Gourmont, Remy de, 105
graft rejection, 183*n*
gravity
and blood circulation and pressure, 3, 74, *76*, 78, 83–84, 179*n*
and physiology, 1–2
grazing, 22
guppies, 147
gustation, *see* taste
gymnasts
results of stressful exercise, 185–186*n*

hallucinations, 124
handball, 178*n*
hangover, 44
Harvey, William, 71–73, 77, 81, 149, 175–178*n*
head trauma
and inability to smell, 113–114
headache
Chinese Restaurant Syndrome, *122*
and dehydration, 44
and hypertension, 86
hearing, 106
elephants, 145
humans, 145
and sensory signals, 112–113
heart
and blood doping, 57
and blood pressure, 81–84
and cardiac output, 15
and discovery of blood circulation, 176*n*
damage to, *67*
enlargement of, 75–76
as excitable tissue, 44
giraffe, 75
mammals, 14–15
percentage of body weight, 161*n*
physiology of, 74–75, *80*
relaxation and pumping phases, 84
role in circulatory system, 71–72
and salt level in blood, 41
signaling process of, 44–45
and stress, 129
and veins, 79
water content, 42
heart attack, 76, 89–90
heart cells
and sodium concentration, 52
heart failure, 38
and congested shock, 89
heart muscle cells, 81–82, 155, 178*n*
heart muscles, 75, 158*n*
death of, 89
heart rate, 81–83
and animal size, 15–16

and body mass, 16
and defense mechanism responses, 127
and emotional distress, 82
of small animals, 82
heart size, 16, 79–81
of athletes, 77, 178*n*
and body size, 15, 75
mountain dwellers, 66
within species, 15
heat
and blood flow, 98
and metabolism, 160*n*
thermal reception, 107
heat absorption
crocodile, *93*
heat dissipation
warm-blooded animals, *100*
heat exchange
and surface area, *100*
heat exchanger, countercurrent, 95
heat loss, *101*
big animals, 13
and body size connection, 16–17
body temperature and surface area, 10–11
bumblebee bat, *12*
and metabolism, 17
minimizing, 99
radiators, 98
and surface area, *101*, 162*n*
heat pits, 113
heat production, 99
body size and metabolic rate, 9–10
brown fat cells, 100–101
glucose conversion, 102
in insects, 96
and metabolic rate, 7–8, 162*n*
shivering, 99
small animals, 11
and thyroid hormone, 101–102
tuna, 95–96
heat radiators, 91–92, 97–98, *101*
and blood flow, 99
dinosaurs, 104
heat retention
large animals, 97
warm-blooded animals, *100, 101*
heat sink, 11
heat stroke, 97
hematocrit, 169*n*
hemoglobin, 56
following hemorrhage, 88
mountain dwellers, 65
and oxygen, 56
in seals, 57–58
hemorrhage, 88, 130
Hench, Philip Showalter, 183*n*
herbivores, 22
herding, 22
hibernation
and body temperature, 103–104
fattening up for, 18, 19
hibernators, 103–104
high blood pressure, *see* hypertension
Himalayas, 69
hippocampus cells
and feedback, 133–135
Hippocrates, 156*n*
Hippocratic oath, 156–157*n*
histamine, 88, 89
homeostasis, 4–5, 90, 127, 132, 156*n*, 160*n*
honeybee, 96
hormones
for shock or low blood pressure, 88
and stress, 128–133
horse
heart size, 15
spleen, 58

humans
  blood circulation, 63–64
  blood pressure, 75, *76*, 88
  body temperature, 92
  brain evolution, 143
  breathing, 14, 55, 70
  calorie consumption, 7
  and capillaries, 47
  cardiac output, 15–16
  and chemical reactions in, 96–97
  diving and blood cell count, 58
  endorphin and pain, 130
  eyes, 118, *119*, 120
  feeding and metabolism, 30–31
  flushed appearance, 98
  focus ability, 118
  future evolution of, 146–147
  hearing, 106, 145
  heart muscles, 178*n*
  heart rate, 15, 81, 82
  heart weight, 16
  improvements on, 144–147
  lung surface area, 46
  metabolic rate, 8, 17
  obesity, 19
  and oxygen deprivation, 55, 60, 68
  physiological evolution of, 142
  salt and water balance, 42, 51–52
  senses, 144
  sensitivity to sweetness, 114, 120–121
  and shock, 86
  stress and, 126, 128, 138–140
  surface area of skin, 98
  taste, *122*
  water elimination, 45
  water locations in, 42
  and water loss and salt, 48
  *see also* man, primitive; men; women
hummingbird, 12
  and earth's magnetic field, 123
  energy burning, 17
  heart rate, 15, 82
  photoreceptors, 116–120
hunger signals, 22–23
hydrocortisone, *see* cortisol
hydrostatic pressure, 61, 181*n*
hyperbaric chamber, 62
hyperglycemia, 135
hypertension, 84–86, 90
  athletes, 77
  causes of, 85
  and cold remedies, 83
  and cortisol, 184*n*
  Cushing's disease, 135
  and heart enlargement, 76–77
  and obesity, 38
  and salt, 85
  treatment for, 85
hyperthermia, *see* overheating
hyperthyroidism, 102
hypertrophy, 76, 77
hyperventilating
  at high altitudes, 66, 169*n*
hyponatremia, 52
hypotension, orthostatic, 84, 86, 90, 179*n*
hypothalamus, 100
  communication network, 25
  feeding-related sites, 23
  functions of, 23–25
  negative feedback, 133
  neuropeptides, 27
  and stress, 128, 129
hypothermia, 139
hypotheses, 150
hypothyroidism, 22
hypovolemia, 169*n*

ice ages
  and primitive man, 139
ice crystals
  formation in blood, 93
icefish, Antarctic, 92–93
IDDM, *see* insulin-dependent diabetes mellitus
iguanas
  salt-pumping cells, 51
immune system, 174–175*n*
  and adrenal hormones, 126
  autoimmune diseases, 33–34
  and cortisol, 136
  diabetes, 34
immunosuppression
  and cortisol, 132, 184*n*
  Cushing's disease, 135
  and glucocorticoids, 135
infants
  blood pressure, 84
  brown fat, 99, 101
  growth of and warmth, 10–11
  mass and surface area, 10–11
infections
  combating stress of, 128
  primitive man, 139
inferior vena cava, 79
inhibitory signals, 113
injuries
  primitive man, 139
inner ear, 106, 182*n*
insects, 147–148
  and attraction to flowers, *116*
  and detecting odor molecules, 114
  heating mechanisms, 96
  sense of smell, 114, 145
  and ultraviolet light detection, 117
insulin
  and diabetes, 32–35
  function of, 165–167*n*
insulin-dependent diabetes mellitus (IDDM), 33–34
intercostal nerves, 174*n*
internal stress response, 125
interstitial fluid, 42, 78
  sodium and potassium levels, 44
intestines
  and circulation, 73
  of fish and salt content of water, 49
  and peptide molecules, 32
  and salt, 51
intracellular fluid, 42
invertebrates
  circulatory systems, 74
  lungs, 168*n*
iron
  in hemoglobin, 56
Italian National Olympic Committee, 77

James I, 176*n*
Jefferson, Thomas, 172*n*
jellyfish, 73–74, 178*n*
jugular veins, 173*n*
juvenile diabetes, *see* insulin-dependent diabetes mellitus (IDDM)

kangaroo rat, 167*n*
Kennedy, John F., 136
kidneys
  and acidic blood, 175*n*
  and adrenal glands, 128
  and aldosterone, 179*n*
  and diabetes, 184*n*
  fish, 50–51
  and vasopressin, 130
Kleiber, Max, 162*n*
Kunz, Thomas, 19

lactic acid, 59–60, 161*n*
lamellae, 168–169*n*

lateral hypothalamic area (LHA), 23, 25
lateral lines (fish), 106–107, 113
Lavoisier, Antoine, 172*n*
leg muscles, 109
leptin, 28–33, 35, 38
lesioning experiments, 23–25
LHA, *see* lateral hypothalamic area
life expectancy
  primitive man, 139
life sciences
  and laws of nature, 1–2
life span
  small animals, 17
light-headedness, 2, 82, 83, 179*n*
light receptors, *see* photoreceptors
limbic system, 128, 129
lions, 6, 22, 114–115
liver
  and defense mechanism responses, 127
  and glucose, 129
  and hemorrhage, 88
liver blood, 71, *72*, 158*n*, 176*n*
llama, 4
lungfish, 46, 167–168*n*
lungs
  air pressure in, 174*n*
  birds, 69–70
  blood pressure in and mountain climbing, 66–67
  book lung, 168*n*
  and circulation, 73, 79, *80*
  and congested shock, 89–90
  mountain dwellers, 65–66
  pulmonary edema, *67*
  seals, 56, 63
  size of, and body size, 14, 163*n*
lupus, 33
lymph tissue
  and stress, 182*n*
lymph vessels, 79

magnetic field, earth's
  and bird orientation, 123
magnetic sensing, 107
malnutrition
  primitive man, 139
Malphighi, Marcello, 177*n*
mammals
  body size and metabolic rate pattern, 9–11
  cardiovascular system, 14–15
  in cold climates, 103–104
  and excess salt, 51
  heart, 14–15, 74–75
  heat dissipation, *100*
  keeping warm, 99
  and low blood pressure, 88
  neocortex, 174*n*
  overheating, 97–98
  oxygen requirements, 55
  and shock, 86
mammals, diving
  and the bends, 62
  blood content, 56, 57
  blood flow in, 64
  and breathing and oxygen, 55–60
  lungs, 63
  and natural blood doping, 58
mammals, small
  bumblebee bat, 12
  hummingbird, 12
  metabolic rate, 7, *9*
  *see also* bats; mouse; shrew
man, primitive
  drive to eat, 21
  and metabolism, 39–40

and starvation, 27, 39–40, 139
and stress, 127, 138–140
taste buds, 121
weight gain as survival mechanism, 20–21
marmots, 39, 103
mass (weight), *see* body mass
matter and energy
and physiology, 1–2
mechanoreception, 106
mechanoreceptors, 109
and sensory cues, 110–115
medulla oblongata, 174*n*
memory
and smell, 144–145
men
blood pressure, 84
breathing rate, 14
calorie consumption, 7
heart rate, 81
metabolic rates, 8
and obesity, 38
sense of smell, 113
Mencken, H. L. , 125
metabolic rate, basal, *see* basal metabolic rate
metabolic rates
and body mass, 16, 161–163*n*
and body size, 9–11, 17
humans, 8, 17
men, 8
shrews and mice, 7–8, *9*
small animals, 16–17
and surface area, 162*n*
tuna, 95
women, 8
metabolic water
recapture of, 167*n*
metabolism
control of, 30–31
and enzymes, 160–161*n*
and fat accumulation, 39–40
fatty rat and obese mice, 26–27
food consumption and body temperature, 11
glucose and fatty acids, 160–161*n*
and heat, 160*n*
and heat generation, 162*n*
and hypothalamus, 23
and oxygen, 160–161*n*
reptiles, 94
slowing down of during dieting, 26–27, 30
mice, *see* mouse
middle ear, 106, 110
migration
of birds and fat stores, 38–39
fattening up for, 18–19
*milieu interieur*, 160*n*
mitochondria, 66, 100, 132
monosodium glutamate (MSG), *122*
Morris, Desmond, 146
moth
and detecting smells, 114
motor control
alcohol and loss of, 181*n*
Mount Everest, 69
mountain climbing, 66–67, 153
oxygen pressure, 169*n*
mountain dwellers, 64–68
physiology of, 65–66
mountain sickness, 66, 67
mouse
body mass, 163*n*
body size and metabolic rate, 10
and eating, 6
heart size, 16
maintaining body temperature, 11
metabolic rate, 7–8, *9*
mutant obese, 26–27, 28
surface area of, 97

mucous membranes
crocodile, *93*
multiple sclerosis, 33
muscle cells
and sodium concentration, 52
muscle contractions
breathing, 174*n*
shivering, 99
muscles
atrophy, 135
and breathing, 54
defense mechanism responses, 127
heart, 75
and lactic acid, 59–60, 161*n*
and oxygen, 58
rib, 174*n*
of seal, 59–60
swimming, 95, 96
of tuna, 95–96
wasting of, 184*n*
myasthenia gravis, 33
myoglobin
seals, 58

nasal mucosa, 109
nasal sprays
and heart rate, 83
National Institutes of Health
on high blood pressure, 84
nature, laws of
and environment, 5
and evolutionary adaptation, 3
and physiology, 1–2
neck
giraffe, 75, *76*
jugular veins, 173*n*
oxygen-detector cells, 54
pulsing in, 172–173*n*
neocortex, 146, 174*n*
nerves, *see* neurons
nerves, intercostal, 174*n*
nerves, phrenic, 174*n*
nesting
of birds and odors, 114
neurobiology
experiments, 24
neuronal signals
and stress, 23, 127
neurons, 164*n*
action potentials, 44
brain, 25–26
and carbon dioxide, 54
and gamma aminobutyric acid, 164*n*
and salt level in blood, 41
and sensory cells, 123
neuropeptide hypothesis, 27
Neuropeptide Y (NPY), 27–31
neuropeptides, 27
neurotransmitters, 164–165*n*
New York City
barometric pressure, 65
Newton, Sir Isaac, 2
NIDDM, *see* non-insulin-dependent diabetes mellitus
nitrogen narcosis, *see* bends, the
nitrogen
in air, 65, 171*n*
and bends, 61–62
Nobel Prizes, 183*n*
non-insulin-dependent diabetes mellitus (NIDDM), 34–35, 166*n*
nonshivering thermogenesis, 99
norepinephrine, 99, 180*n*
nose
blood vessels in, 184*n*
receptors in, 110
nostrils, 2–3, 113
"noxious agents," *see* stress
NPY, *see* Neuropeptide Y
nuclei (in brain), 164*n*

and communication, 25
and eating, 22, 23
in hypothalamus, 23
nursing
and stimuli, 109
nutrients
and capillaries, 78
and circulation, 73, *80*
and digestive system of small animals, 13

obese gene, 29
obese mutant mice, 26–28, 35
obese rodents
and leptin, 29
obesity, 19, 35–36
and associated diseases, 38
and body mass index, 36–38
chronic, 36
Cushing's disease, *137*
and diabetes, 167*n*
domesticated animals, 20
early man, 20
and leptin, 28, 32–33
and Neuropeptide Y, 27
set point for, 31
as survival mechanism, 20–21
in United States, 36, 38
"obesity" gene, 21
octopus, 74
odor molecules
amplification of, 112
detection of, 114
odor receptors, 15, 111
odors
detection by snakes, 113
and nostrils, 113
and sensory cells, 123
and smell center in rodents, 4
and stimuli, 107
*see also* smell
olfaction, *see* smell
operculum, 46, 169*n*
oral contraceptive, 133
orientation
of birds and earth's magnetic field, 123
orthostatic hypotension, 84, 86, 90, 179*n*
Osler, Sir William, 184*n*
osmosis (water movement)
in fish, 47–49
in humans, 42, *43*
osteoporosis
primitive man, 139
ostrich
eyes, 118, *119*
overeating
by children and growth of fat cells, 29
and feeding center, 25
and Neuropeptide Y, 27
and ventromedial nuclei, 23
overheating (hyperthermia)
preventing, 97–98
temperature detectors, 98
warm-blooded animals, 97
ovulation, 133
owls
ears, 145
vision and eyes, *119*
oxygen, 5, 53–70
in air, 65, 171*n*
in alveoli, *67*
and blood circulation, 4
and breathing rate, 14
and capillaries, 78
and chemoreceptors, 106
and circulatory system, 79, *80*
and congested shock, 90
discovery of, 170–171*n*, 172*n*
and exercise, 59
and food intake, 14
and hemoglobin, 56
at high altitudes, 64–68

maintaining constant levels of, 4
and metabolism, 160–161*n*
and survival, vii–viii, 4
transport of, vii, 4
in water, 47
oxygen capture
at high altitudes, 65
by fish, 45–48
by jellyfish, 74
by mountain dwellers, 66
oxygen consumption
and body size, 9–10
and metabolic rate, 8
shrews, 14
oxygen deprivation, 60, 68, 128
oxygen-detector cells, 54
oxygen pressure
in blood, 68
and mountain climbing, 169*n*
oxygen reservoirs
and myoglobin, 58

pacemaker, artificial, 82
pacemaker, natural, *see* sinoatrial node
pacemaker cells, 82
pacemaker nerves, 178*n*
pain
combating stress of, 128
endorphin, *131*
signals, 123, 124, 129
painkillers
endorphin, 130
pairs
of physiological structures, 3, 113
pancreas
and diabetes, 33–35
parabiosis experiments, 28
Parkinson's disease, 24
peptides
and control of feeding and metabolism, 30–31
leptin, 28–31
Neuropeptide Y, 27–31
*see also* neuropeptides
perception, sensory, 115–116
peregrine falcon
sight, 117–118
pernicious anemia, 22, 138
perspiration
and body water levels, 170*n*
and salt levels, 48
Peruvian mountain dwellers, 65–66
pH balance
of blood, 59, 175*n*
pharynx
taste buds in, 182*n*
photoreceptors
birds, 116–120
phrenic nerve, 174*n*
physical laws
and physiology, 1–2
*physiologoi*, 156*n*
physiology, 2, 156*n*
experimental study of, 149–155
and laws of nature, 1–2
of mountain dwellers, 65–66
nature of, vii
origins, 156–159*n*
and physical laws, 1–2
scaling, 161*n*
pigeons
and earth's magnetic field, 123
vision and eyes, *119*
pigs
neocortex, 146
and taste buds, 120
pill, the, 133
pilots
and the bends, 62
pituitary gland, 23, 100
adrenocorticotropic hormone, 129

and Cushing's disease, 184*n*
endorphins, *131*
and feedback, 133
tumors of, 135, *137*
pituitary hormones
in women, 132–133
planarian
sensory stimuli, 107–108
plants
oxygen production, 171*n*
plasma, blood, 81
carbon dioxide in, 54
following hemorrhage, 88
salt level of, 42
poisonous foods, 121
pons, 174*n*
positional sense, 106
potassium, 44
balance with sodium and water, 45
high level in blood, 136
predators, 94
eyes, *119*, 120
prednisone, *137*
pressure
negative, in fish, 46
relationship with volume, 168*n*
*see also* barometric pressure; blood pressure; diastolic pressure; hydrostatic pressure; systolic pressure; water pressure
pressure receptors, 106
Priestley, Joseph, 53,170–171*n*
Priestley crater (Mars), 172*n*
primitive man, *see* man, primitive
proprioceptors, 109, 181*n*
protein receptor molecule, 110, 111
puberty
gymnasts, 185–186*n*
pulmonary edema, *67*, 68
pulse
in neck, 172–173*n*
pumps
circulatory systems, *74–77*
molecular, 49, 50
pyruvic acid, 60

rat
mutant fatty, 26–27
sensory cues and sight, 4
Reade, W. Winwood, 142
receptor cells, 54, 55
and mechanoreception, 106
and odor, 106
sensitivity of, 115
taste, 106
*see also* baroreceptors; chemoreceptors
red blood cells, 81
and blood doping, 56–57
and carbon dioxide, 55
death of, *43*
hematocrit, 169*n*
and hemorrhage, 88
mountain dwellers, 56–57, 65–66
osmotic movement of water, *43*
oxygen and, 56–57
production of, 56
in seals, 57–58
and spleen, 175*n*
reflexes
and sensory stimuli, 107–109
renin, 179*n*
reproductive glands
and stress, 182*n*
reproductive hormones
and olfactory sensitivity, 114
reproductive system
and stressful exercise, 185*n*
reptiles, 94–96

body temperature, 92
breathing, 55
cortex, 174*n*
evolution, 147
heat and metabolism, 94, 95
heat absorption, *93*
responses to stress, 127
salt-pumping cells, 51
thermal receptors, 113
toleration of freezing, 92
respiration, 174*n*
anaerobic, 59–60, 161*n*
oxygen consumption, 8
respirator, 174*n*
respiratory center
and oxygen detection, 54
respiratory diseases, 55
respiratory systems
birds, 69–70
mountain dwellers, 66
retina
birds, 118, 120
seal, 60
Rockefeller University, 36
rodents
brown fat, 99
life span, 17
and sensory input, 4, 108
smell center, 4
torpor, 103
rowing, 178*n*
Rubner, Max, 162*n*

saccharine, 120
salt
in blood, *43*
and blood water loss, 48
and body water content, 45
in fish, 48–50
and hypertension, 85
levels in blood, 41–42, 51, 52, 184*n*
maintaining constant levels of, 4, 5
and perspiration, 48, 170*n*
seawater level of and fish, 2
and thirst, 41
and water balance, 51, 183*n*
salt taste, *122*
salt taste buds, 120, 121
salt-detecting cells, 51
salt-pumping cells, 51
satiety centers, 24, 25
saturated fats
in fish, 103
scaling (physiology), 161–163*n*
scurvy, 22
sea level
air pressure, 65
seagulls
salt excretion, 51
seal
adrenal glands, 60
and anaerobic respiration, 59
and the bends, 2, 62
blood supply, 63
diving and breathing, 55–59, 62–63
fetus, 60
and hemoglobin, 57–58
lungs, 56, 63
muscles, 59–60
and oxygen deprivation, 68
seawater
drinking, 48–49
seaweed, *122*
seeing, 115–120
*see also* eyes; sight; vision
seizure, 184*n*
Selye, Hans, 125, 138, 182–183*n*
senses, 105–124
evolution of, 1097
*see also* hearing; sight; smell, sense of; taste

sensory cells, 110
  and action potentials, 123–124
  adaptations, 115
  and amplification of odors, 112
sensory cues, 106, 108
  and mechanisms for responses, 110–115
sensory perception, 115–116
  of bird eyes, 118
sensory stimuli, 106–107
  amplification of, 112
  detecting odors, 114
  detection and interpretation of, 123
  and reflexes, 107–109
Serengeti plains, *131*
series circuit
  circulatory system, *80*
Serveto, Miguel (Michael Servetus), 177*n*
set point (for body weight), 31
  and fat accumulation, 39–40
sex hormones
  and stressful exercise, 185*n*
shark
  and electrical signals, 2, 123
  and musculature, 95
  senses, 106
  sensory cues and sight, 4
shellfish
  allergic reactions to, 88
Sherpas, 147
shivering, 99
shock, 60
  anaphylactic, 88–89
  circulatory, 86–90, 128
  congested, 89–90
shrew, 5
  cardiac output, 15–16
  circulatory system, 16
  digestive system, 13
  food intake, 11–14, 97
  heart rate, 15
  maintaining body temperature, 11
  metabolic rate, 7, *9*
  oxygen consumption, 14
sight
  birds, 106, 114, *119*
  and environment, 106
sinoatrial node, 81–82
size, *see* body size
skiing, alpine, 178*n*
skin
  and mechanoreception, 106
  thinning of and cortisol, 184*n*
smell, sense of, 113–114
  birds, 114
  insects, 145
  loss of, 113–114
  and mechanoreceptors, 109
  and memory, 144–145
  nocturnal animals, 106
  and receptor cells, 110, 114–115
  and stimuli, 107
  *see also* odors
smell center
  rodents, 4
smoking, 86
snake
  forked tongue and odors, 113
  and heat and metabolism, 95
sodium, 44
  balance with potassium and water, 45
  and chemical receptors, 106
  low level in blood, 136
sodium chloride, *see* salt
sodium-potassium molecular pump, 101–102

soft drinks
carbonation, 61, 63
and sodium, 41
sound
hearing, 106
location of source of, 112
low-frequency, and elephants, 114
nocturnal animals, 106
sound stimuli, 109
sound waves, 110
sour/acid taste, *122*
sour taste buds, 121
species classification, 156*n*
species comparisons
metabolic rates, 8
spiders
lungs, 168*n*
spleen, 57–58, 174–175*n*
seal, 60
sports
blood doping, 174*n*
Gatorade, 169–170*n*
performance enhancing, 56–57
*see also* athletes
sports drinks, 48, 169–170*n*
squid, 74
squirrels
and eating behavior, 6, 7
fat stores and hibernation, 39
hibernation, 103
starlings, 114
starvation
combating stress of, 128
primitive man, 39–40, 139
prolonged, and adaptation to, 27
slowdown of metabolism, 30
stegosaurus, 104
steroid hormones, 138, 165*n*
anabolic, 57
and cholesterol, 129, 183*n*
synthetic adrenal, *137*
stimulation experiments (of brain), 25
stimuli
stressful, 126
two sets of receptors for, 113
stimulus molecule
and protein receptor molecule, 110
stress
chronic, 133
early observations of, 182*n*
hormones, 130–132, 141, 165*n*
and internal changes, 182*n*
physical, 128
primitive man, 138–140
stress research, 183*n*
stress responses, 125–131
evolution of, 141
and natural painkiller, 130
zebra, *131*
stressful inputs, 129
stretch receptors, 106, 109
stroke, 38, 60, 174*n*
and hypertension, 86
sucrose
and taste buds, 120–121
suffocation
fish, 169*n*
sugar
sensitivity to, 114
and taste buds, 120
*see also* glucose; hyperglycemia
sugar, table (sucrose)
and taste buds, 120–121
suicide
assisted, 156*n*, 157*n*
by wrist slicing, 172*n*
superior vena cava, 79

surface area, 91–92
  of bat wings, 91, 98
  and dinosaurs, 104
  of elephant ears, 98
  of gills, 46
  and heat loss, 162*n*
  and heat retention, 97
  and mass, 10–11
  and metabolic rate, 162*n*
  and metabolism, 17
  and warmth, 10–11, 99
surface area–body mass relationship, 91, 162–163*n*
surface area–to–weight ratio, 10–11, *12*
Surface Rule hypothesis, 162*n*, 163*n*
survival adaptations
  taste buds, 121
survival mechanisms
  weight gain, 20–21
survival strategies, 4, 5
sweet taste buds, 120
sweeteners, artificial
  detecting, 111, 120
sweetness
  sensitivity to, 120–121, *122*
sympathetics, 179–180*n*
systolic pressure, 84

Tae kwon do, 178*n*
taste
  and loss of sense of smell, 114
  and poisonous foods, 121
  receptors, 111
  and stimuli, 107
taste buds, 120–123, 182*n*
tea
  and heart rate, 83
temperature, *see* body temperature
temperature detectors, 98
testosterone, 183*n*
theophylline, 83
thermal receptors, 107
  heat pits, 113
  reptiles, 113
thermodynamics
  and physiology, 1–2
thermostats, *see* temperature detectors
thirst, 51
  and angiotensin II, 179*n*
  from blood loss, 87
  diabetes, 165*n*, 166*n*
  and salt, 41
throat, *see* pharynx
thyroid gland, 23, 100
thyroid hormone, 100–102
tissues
  and excess salt, 44
tongue, *122*
tooth decay
  primitive man, 139
torpor, 103
touch
  and mechanoreceptors, 106
trachea
  birds, 70
transplant rejection
  and steroids, *137*
transplantations, 60
triglycerides, 29
trout, 46
tuberculosis, 138
tumors
  and cerebellar function, 181*n*
  Cushing's disease, 184–185*n*
  pituitary gland, 135, *137*
tuna, 95–96
*Tyrannosaurus rex* , 104

ultraviolet light
  birds and photoreceptors, 116–118
United States
  cardiovascular diseases, 85

overweight population, 36, 38
U.S. Department of Health and Human Services, 85
urination
and diabetes, 165*n*, 166*n*
urine
glucose in, 166*n*

vasopressin, 130, 183–184*n*
vaso-vagal syncope, 82
vegetative functions, 174*n*
veins, 14, 78, 79
and blood pressure, 79, 172*n*
circulatory system, *80*
tuna, 95
vena cava, inferior and superior, 79
ventricles, 75–77, 79, *80*, 173*n*
damage to, *67*
and heart attack, 89–90
thickness of, 77
ventricular hypertrophy, 76, 77
ventromedial nuclei (VMN), 23–24
vertebrates
brain stem evolution, 174*n*
heart, 74–75
vertigo, 182*n*
vibrational signals, 113
vibrations
mechanoreceptors, 106
vision
and diabetes, 165*n*
photons, 107
*see also* eyes; seeing; sight
vision, field of, *119*
and lateral-facing eyes, 120
visual cortex, 124
visual cues
stressful inputs, 129
vitamin B12, 138
vitamin C, 121
vitamin D
and cholesterol, 183*n*
VMN, *see* ventromedial nuclei
volume
relationship with pressure, 168*n*

warm-blooded animals, *see* animals, warm-blooded
warm-bodied animals, 96
waste products, 163*n*
lactic acid, 59–60, 161*n*
water, 45
water
balance with sodium and potassium, 45
in blood, 48
in body and salt level, 45
and desert animals, 45
detecting taste of, 120, *122*
and diabetes, 184*n*
excessive consumption of, 167*n*
levels in body and results, 41–44
locations in body, 42
oxygen capture from by fish, 45–48
oxygen content, 47
and salt balance, 51, 183*n*
and salt content of blood, 41–42
salt content and fish, 2
and survival, 4
as waste product, 45
*see also* osmosis
water breathing, 47, 48
water intoxication, 167*n*
water loss
in blood, 169*n*

and blood pressure, 47–48
in fish, 47, 49
perspiration, 48
water polo, 178*n*
water pressure, 65
in fish, 46
water temperature
and fish body temperature, 102, 103
water vapor
and elimination of water, 45
Wedell seal, 56
weight, *see* body mass
weight gain
mutant fatty rat and obese mouse, 26–27
weight loss
and diabetes, 167*n*
weight reduction
and leptin, 35
whales
blue, 16
diving depths, 62
and diving underwater, 56, 63
white blood cells, 81, 174*n*
and bacteria, 106
women
blood pressure, 84
breathing rate, 14
calorie consumption, 7
heart rate, 81
metabolic rates, 8
and obesity, 38
and positive and negative feedback, 132–133
sense of smell, 113
wood frog, 92

yo-yo effect (dieting), 26, 30

zebra
stress response, *131*